© NASA, ESA, CSA, STScI

Welcome

Astrophysics is one of the most fascinating fields of science – studying the many wonders of the universe can help us better understand our place in it. In this brand-new special edition, we've compiled a selection of articles from the *All About Space*, *How It Works*, and *Space.com* archives to explain the science behind cosmic questions, including how galaxies form and what distant stars and planets are made of. We will also investigate some of the biggest mysteries in the cosmos, such as how the universe began and whether there is life beyond Earth.

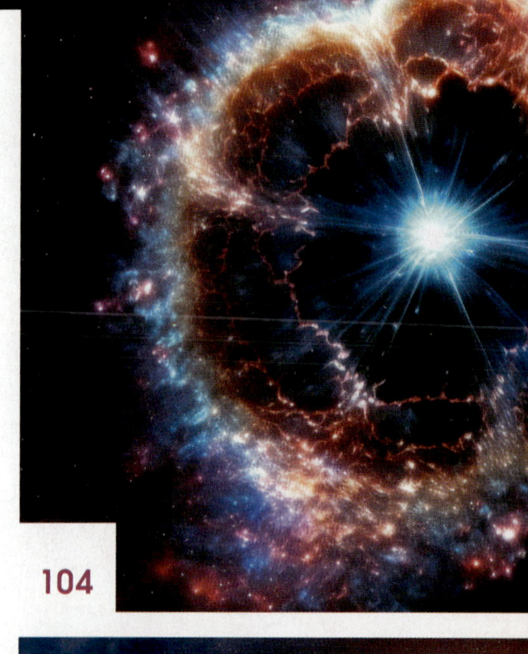

Contents

06 Complete guide to the universe
From how it all began to its ultimate fate

14 The Big Bang theory
It's our best model of how the universe works, but where did it come from?

20 What happened before the Big Bang?
Could there have been a time before the birth of the universe?

28 What is a light year?
How we measure the vast distances across the universe

30 How to build a galaxy
The making of these billion-star structures has been puzzling astronomers for decades

38 Secrets of black holes
Places where the laws of physics are pushed to the extreme

48 Everything you need to know about the Solar System
Take a tour through our current understanding of the planetary system we call home

54 Does the Sun stay still?
We may think of it as the stationary centre of our Solar System, but the planets make our Sun wobble

56 Complete guide to exoplanets
Our knowledge of worlds beyond the Solar System has exploded in the last three decades

64 How many stars are there in the universe?
Will it ever be possible to come to an accurate estimate?

70 Finding and defining exomoons
As our catalogue of exoplanets grows, the hunt is on for moons around distant worlds

104

48

14

122

56

38

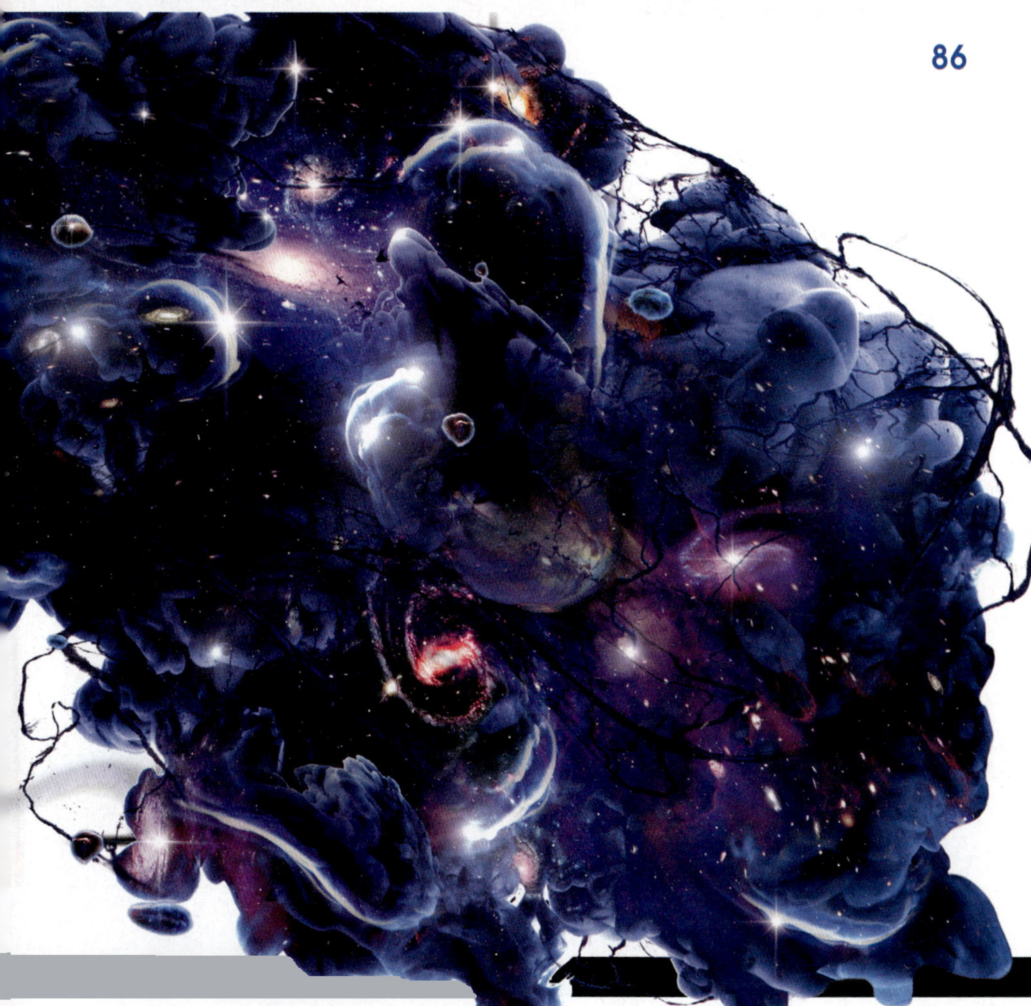

86

78 What are planets like on the inside?
Even among the worlds of our Solar System, we see a huge variety of planets

80 Extreme stars
From the biggest and brightest to the smallest and dimmest, discover some stellar extremes

86 Dark matter: where did it come from?
How bubbles could be the answer to the universe's most mysterious substance

94 How did Earth get its water?
Moon rocks suggest that the water might have been here all along

96 Celebrating Hubble
The Hubble Space Telescope has astounded for decades, returning breathtaking images and groundbreaking science

104 When stars go supernova
The titanic stellar eruptions that give rise to life

112 Dark energy
It's the most mystifying phenomenon in the universe: the force tearing space apart

120 Why are we still searching for intelligent alien life?
Should we widen our search criteria when looking for extraterrestrials?

122 Are we living in a multiverse?
Could our entire universe be just one small part of many?

Complete guide to the universse

...from how it began to its ultimate fate

© Adrian Mann; Getty

Our universe encompasses everything around us, countless stars, planets and galaxies stretching as far as our telescopes can see, with vast expanses beyond them that lie forever out of sight. It's a big subject – the biggest of all – so join us as we break down the basics

What is the universe?

The universe is a vast expanse of space and time filled with matter and energy. The word comes from the Latin universus, meaning 'all things', although ours may in fact be just one of many universes in a so-called multiverse. Over the course of billions of years, the simple atoms that formed in the early history of the universe have come together to produce countless galaxies, stars and solar systems like our own, separated by enormous distances. Meanwhile, nuclear processes inside stars, or triggered by their violent deaths, have steadily processed the first simple atoms to enrich later generations of stars, and the space between them, with heavier and more complex elements in an ongoing cosmic recycling scheme.

Cosmic distances

Astronomers put a scale to the universe by measuring the distance to remote galaxies. The most accurate way of doing this is to look for 'standard candles' – rare objects that are bright enough to see across vast distances and which obey physical rules that reveal their intrinsic luminosity or light output. The most useful standard candles are variable stars called Cepheids, whose pulsation period is linked to their average luminosity. The distance to the host galaxies of these stars can be worked out by comparing their theoretical luminosity to their brightness measured on Earth.

While Cepheids are only detectable in the relatively local universe, the distance measurements they provide are enough to reveal a rule of thumb that can be applied more widely. This rule, known as Hubble's law, links the distance of a remote galaxy to the speed at which it's moving away from us. Hubble's law applies because the universe as a whole is expanding, carrying distant galaxies away from each other like raisins in a rising cake. The wider the separation between galaxies, the faster they move apart, because there is more expanding 'dough' between them. What's more, this expansion stretches light from fast-retreating galaxies into longer and redder wavelengths, creating a 'redshift' that can be measured on Earth and used as a proxy for distance itself. Broadly speaking, the farther away a galaxy lies, the greater the redshift in its light.

Expansion and origins

Since the universe as a whole is expanding, it must have been much smaller in the past, with matter packed more closely together. Modern measurements of the rate of expansion – known as the Hubble constant – put it at around 21.5 kilometres (13.4 miles) per second per million light years of separation; in other words, on average galaxies 10 million light years apart are receding from each other at around 215 kilometres (134 miles) per second. By winding back the clock of cosmic expansion to zero, we can pin down a time when everything was in the same place: the beginning of the universe, about 13.8 billion years ago. Because matter was packed together more closely in the early universe, it must also have been hotter. In the very earliest times, temperatures were so high that any matter that formed would have fallen apart – in effect, matter and energy were interchangeable in line with Einstein's famous equation $E=mc^2$. However, as temperatures cooled, more complex forms of matter were able to persist for longer, eventually forming stable atoms of simple elements such as hydrogen and helium. This is the origin of the Big Bang theory – our best explanation for how the universe began.

Cosmic time machine

An intrinsic property of the universe is that the speed of light in a vacuum is constant. Light moves at 299,792 kilometres (186,282 miles) per second – a speed that seems instantaneous in everyday life, but which is nevertheless finite – and this has important consequences for how we view distant space. In essence, the further away we look in space, the further we are looking back in time, since light must have left objects long ago to reach our telescopes now. The evolution of stars and galaxies is so slow that the millions of years taken by light from

© Adrian Mann

Life in the universe

One of the most remarkable aspects of the universe is that it's given rise – at least once – to intelligent life. Life of any kind is only possible because of a handful of basic parameters of nature, such as those which control the strength of forces within and between atoms and those that permit the existence of a stable universe at all. The precise values of these parameters allow the creation of stable elements, the existence of habitable planets and the formation of complex molecules.

Instinctively it may feel like this level of 'fine-tuning' is too big a stroke of good fortune to happen by chance, but the so-called anthropic (human-centred) principle points out a possible flaw in this logic: we shouldn't be too surprised that the universe seems fine-tuned for our existence because if its properties were slightly different, we simply wouldn't be around to measure them.

A more immediate question, however, is whether life elsewhere in the cosmos is rare or widespread? Breakthroughs in recent decades, such as the discovery of abundant exoplanets orbiting other stars, have shown that raw materials and suitable habitats for life are certainly widespread, but we still can't be sure that life will always arise wherever conditions are broadly hospitable or whether we're overlooking some vital factor.

An artist's concept of a hypothetical planet orbiting in the habitable zone of a red dwarf

© D. Aguilar/Harvard-Smithsonian Center for Astrophysics

Exploring the multiverse

Many cosmologists believe that our universe is just one within a wider 'multiverse'. Many complex types of multiverse are possible – for instance, a theory called eternal inflation argues that just as our universe arose from one tiny region within the original Big Bang and was then blown up by a sudden event called inflation, the raw material of the Big Bang is still out there, giving rise to new universes with the potential for very different physical parameters to ours.

Another is closer to science fiction's 'parallel' universes, and is rooted in quantum mechanics, the bizarre branch of physics that governs particles on a subatomic scale, within which events always carry a degree of uncertainty. There are many ways of handling this uncertainty, but according to one unusual idea, the 'many-worlds interpretation', every possible outcome of a quantum event occurs, but each one happens in its own distinct universe. In this sense, ours may be just one among a truly infinite number within which things happen to have played out a certain way.

Voyager 2 is over 17.5 billion kilometres (11 billion miles) from Earth, and will continue its journey out into the universe for years to come

nearby galaxies makes little difference, but as telescopes and detectors have improved, it's become possible to see objects that are many billions of light years away – so distant that we are looking back to a much earlier phase of cosmic history, when they were just beginning to form.

The first objects to be spotted at these immense distances were quasars – galaxies with monster black holes at their centres that have not yet settled down into a quiet slumber, still actively feeding on gas and dust from their surroundings. Intense light and other radiation from their central regions gives them a star-like appearance, and the surrounding galaxies only became visible as images improved. Telescopes such as the Hubble Space Telescope, meanwhile, have allowed astronomers to detect the light from fainter galaxies even closer to the Big Bang, revealing an era of chaotic mergers as small galaxies grew into larger ones such as today's spiral and elliptical systems.

The edge of the universe

The finite speed of light puts a barrier around 'our' universe that we cannot penetrate. Since radiation has only had 13.8 billion years to reach Earth, it's impossible for any information to reach us from remote parts of the universe beyond 13.8 billion light years. This defines our 'observable universe', a vast bubble of space centred on Earth. Objects at the very edge of this bubble are still too faint and distant for our telescopes to capture. Before the first galaxies, an early generation of truly monstrous stars were the first luminous objects to light up the young universe, and it's hoped that some of these may finally come into view with the launch of the giant James Webb Space Telescope

"More complex forms of matter were able to persist"

Elusive dark matter can sometimes be mapped due to the effects of its gravity. In this example, dark matter (blue) has separated from hot gas (pink) during the collision between a pair of distant galaxy clusters

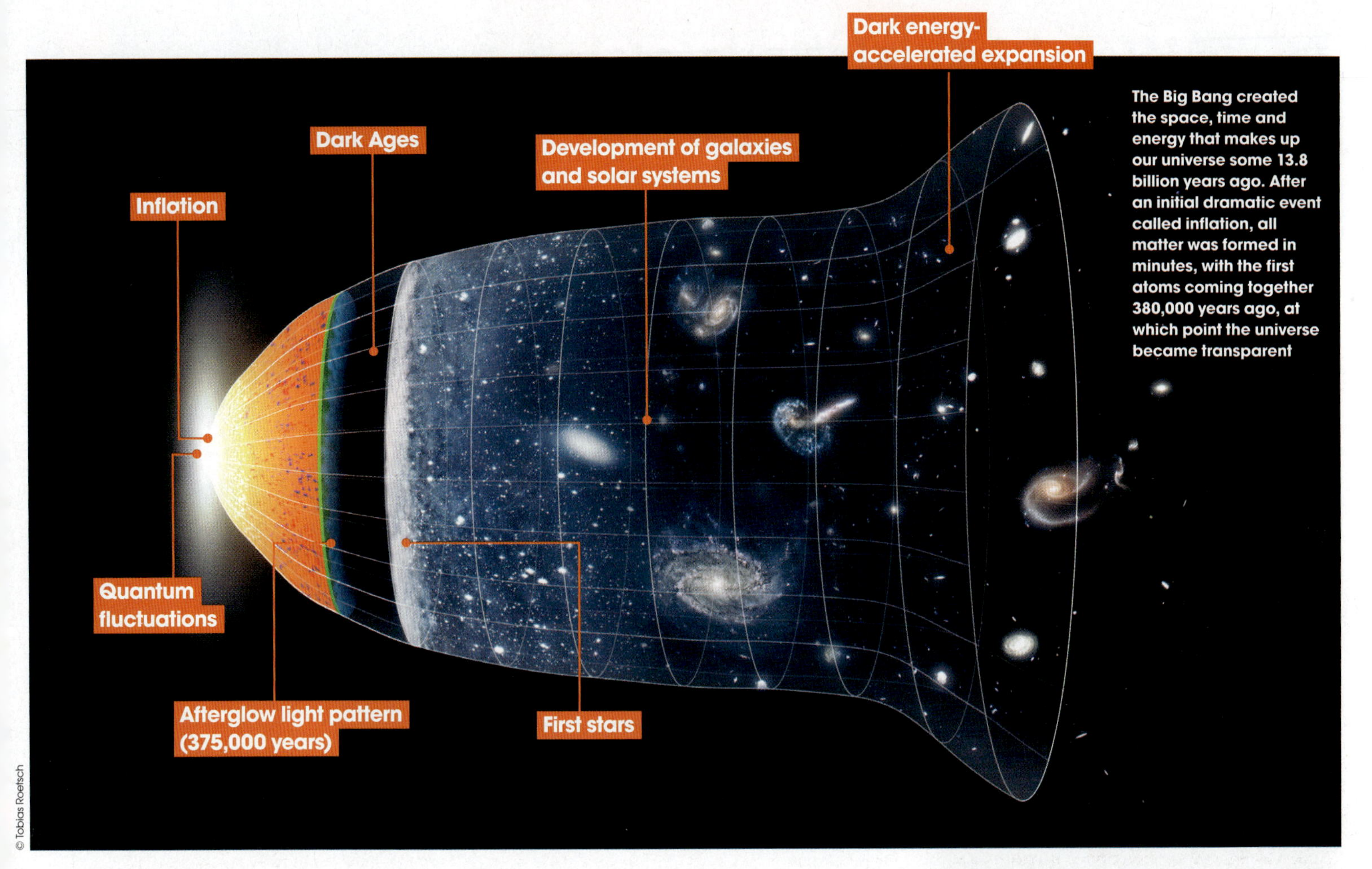

The Big Bang created the space, time and energy that makes up our universe some 13.8 billion years ago. After an initial dramatic event called inflation, all matter was formed in minutes, with the first atoms coming together 380,000 years ago, at which point the universe became transparent

later this year. Beyond these lies the 'cosmic dark age', a period of around 400 million years after the incandescent Big Bang had cooled into invisibility, during which matter coalesced in darkness.

There's another reason for the darkness at the very edge of our observable universe. As the speed at which distant regions and objects are retreating from us gets ever more extreme in all directions, light is stretched and redshifted beyond the limits of visibility – first into the invisible infrared part of the spectrum, and then into microwaves, which are short-wavelength radio waves. The most distant radiation of all comes from the cosmic microwave background (CMB) – a faint signal that comes from all over the sky. It's the afterglow of radiation that escaped from the incandescent fog of matter and trapped light when the universe finally cooled enough to become transparent, about 380,000 years after the Big Bang itself.

Matter and forces

The way that matter is spread across the universe is controlled by four fundamental forces, which influence different types of matter in different ways. Visible matter is composed of tiny atoms that are bound to each other by the electromagnetic force. These are built from even smaller elementary particles held together by both electromagnetism and two 'nuclear forces' that are much more powerful but only operate across tiny distances.

When matter accumulates in large quantities, another force, gravitation, or gravity, comes into play. Gravitation attracts objects with mass. It's extremely weak between individual atoms, but grows stronger as they accumulate in bulk, and has an effectively limitless range. By drawing material together on a variety of scales, gravity is responsible for the formation of stars, galaxies and even galaxy clusters. However, there has not been enough time since the universe was born for it to have shaped the very largest structures in the universe: billion-light-year chains or filaments of galaxy clusters and superclusters that surround apparently empty voids. This large-scale distribution of matter originates instead from density variations that have been with the universe since birth. Such 'ripples' can be seen as tiny differences in the wavelengths of the CMB radiation released as the universe became transparent.

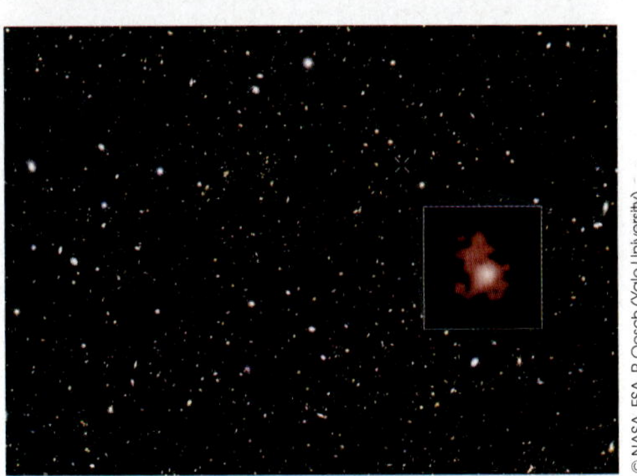

The most distant galaxy so far detected, known as GN-z11, exists about 400 million years after the Big Bang itself

Dark matter and dark energy

The visible matter that produces and interacts with light and other forms of radiation amounts to just five per cent of all the matter in the universe. The remainder, 'dark matter', is not only dark, but also transparent and completely unaffected by radiation. It gives

Bigger and bigger

The scale of the universe is so vast that it can be hard to grasp. One of the best ways to comprehend it is by starting on a relatively small scale with our home planet, Earth, and working outwards

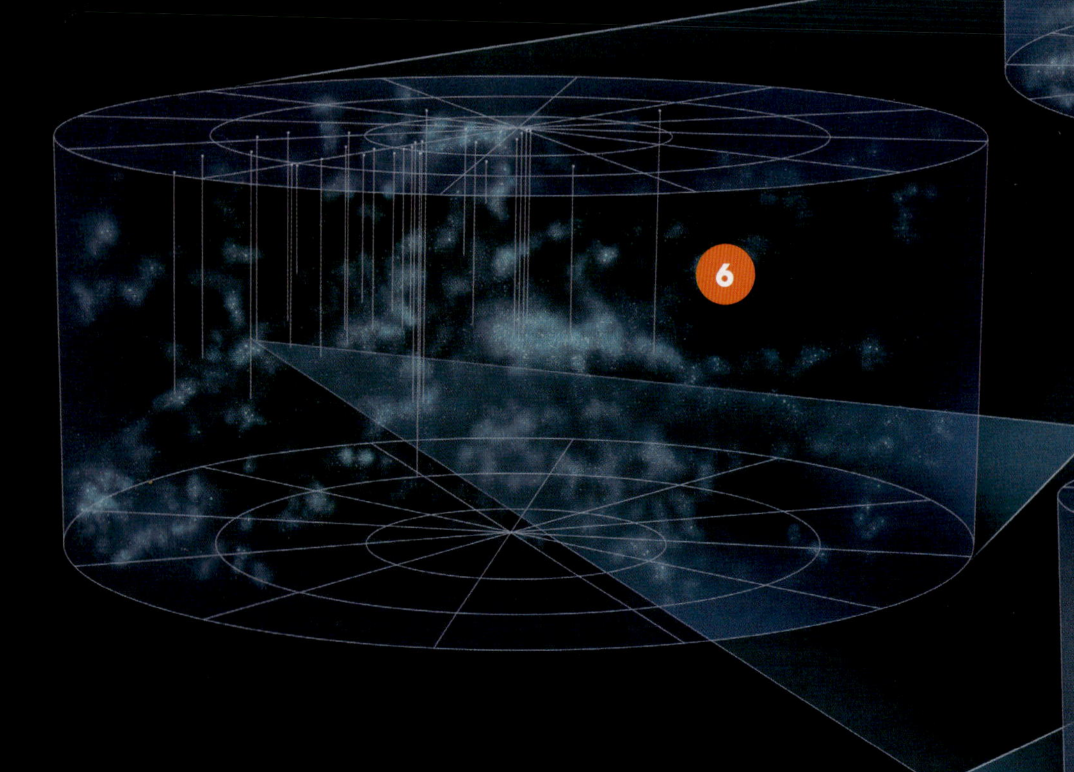

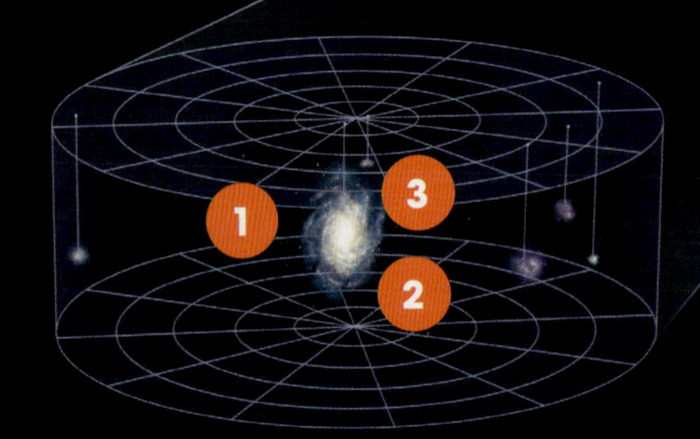

"The universe as a whole is expanding"

1 Earth Earth's diameter is 12,756 kilometres (7,926 miles). The Moon orbits Earth at an average of 384,400 kilometres (238,855 miles).

2 Solar System The outermost planet in the Solar System, Neptune, orbits the Sun at a distance of 4.5 billion kilometres (2.8 billion miles).

3 Living local Stars in our local region of space are separated by light years – tens of trillions of kilometres. The brightest star in the sky, Sirius, is 8.6 light years away.

4 Home galaxy Our Sun and Solar System, and all the stars in our sky, are members of the Milky Way – a vast spiral of stars that's roughly 120,000 light years across.

5 Grouping up The Milky Way is a major member of a small galaxy cluster called the Local Group, occupying a volume of space about 10 million light years across.

6 Big cluster The Local Group is an outlying region of our local galaxy supercluster, sometimes called the Virgo or Laniakea Supercluster. It is over 100 million light years long.

7 Empty space The Virgo Supercluster is part of a local supercluster complex a billion light years across. At this level, the large-scale structure of filaments and empty voids begins to emerge.

Universe by numbers

200 billion

Estimated number of galaxies in the observable universe

9.46 trillion kilometres

A light year, the distance light travels in one Earth year

46.5 billion light years

The distance of an object at the limit of our observable universe

2.726k

The measured temperature of CMB radiation in Kelvin

21.5km/s per million light years

The average rate at which objects move away from each other in the universe due to the expansion of space

itself away only through its gravitational influence, altering the rate at which galaxies rotate and affecting the movement of galaxies within clusters. The nature of dark matter is a mystery, but astronomers have ruled out some possibilities.

We can be pretty sure that it's not just made up of small, dark objects that are undetectable through current instruments – for example stray planets, failed stars or black holes. Studies show that there simply aren't enough of these to make much difference. Meanwhile, models of the matter's distribution based on its gravitational effects show that it clumps loosely around visible objects. It's not spread uniformly across the universe and it's not an explanation for the relatively empty voids of space in the large-scale cosmos. The current best guess is that dark matter is predominantly made from one or more unknown types of elementary particles, known as weakly interacting massive particles (WIMPs). This name says very little about what the particles actually are, and standard models of particle physics do little to address the question, but it seems that answers are most likely to emerge from 'new physics' discovered in experiments such as the Large Hadron Collider than through direct astronomical observations.

Meanwhile, there's another even more elusive and troubling 'substance' in the universe.

The cosmic microwave background marks the edge of our observable universe – a wall of light red-shifted into microwaves from the moment that the universe first became transparent, about 380,000 years after the Big Bang

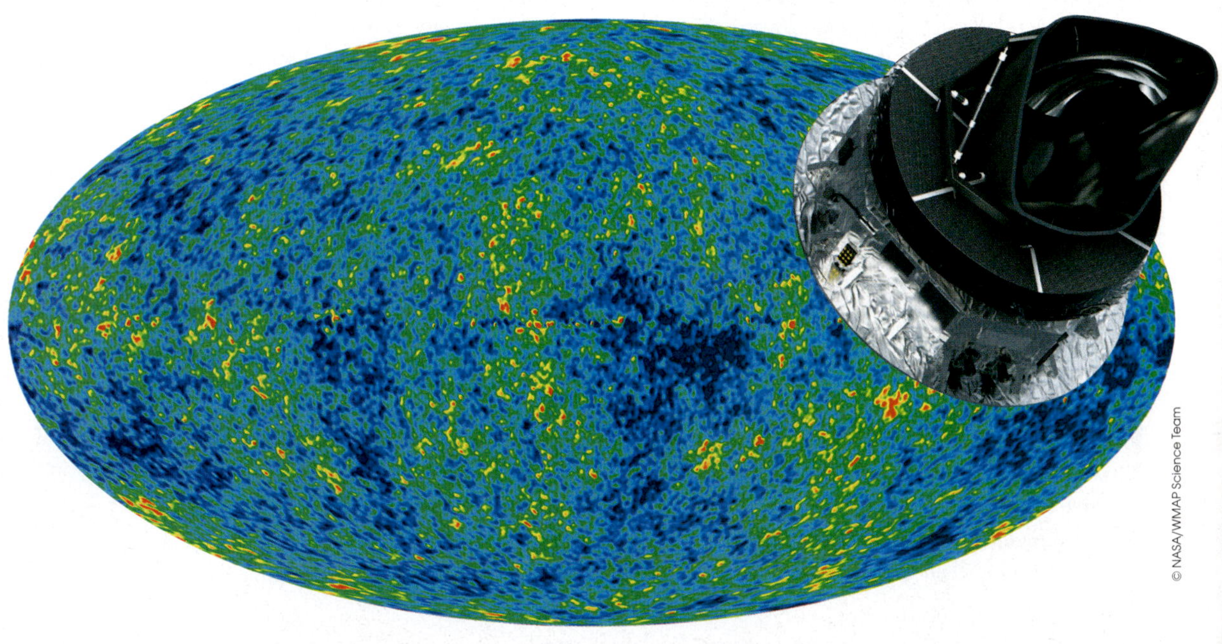

8 Far-reaching The observable universe has a diameter of about 93 billion light years based on the current locations of regions that we can see.

Large clusters like Abell 2744 are the biggest objects governed entirely by gravity

The presence of so-called 'dark energy' was discovered in the late 1990s when astronomers found evidence that cosmic expansion is currently accelerating, rather than slowing down as we might expect due to the gravity of all the matter in the universe. Calculations show that it must account for around 71 per cent of all the energy in the universe. The nature of dark energy remains puzzling, and some still say the evidence for it is inconclusive. Among those who support the theory there are two popular explanations: it may be either an intrinsic property of space itself, or a 'fifth force' that acts like a form of antigravity on very large scales.

Fate of the universe

While 13.8 billion years is a long time, the universe is still young in some respects – some stars shining today formed alongside the first galaxies, and even our Sun is more than one-third the age of the universe. So what will happen to it in the future? On the largest scales, the fate of the cosmos is determined by the balance between its tendency to expand and the inward pull of gravity from all the matter it contains. Astronomers used to assume cosmic expansion would naturally slow down over time. This could lead to a universe that grew ever more slowly, with an effectively infinite life span during which stars would gradually process all of their available materials until they no longer had fuel to shine, and even matter itself might eventually disintegrate. In contrast to this 'Big Chill' scenario, gravity might win out and reverse the expansion of space, drawing everything back into a hot, dense 'Big Crunch'. For a long time, estimates of the density of matter in the universe placed it on the fence between these two possible fates.

But the discovery of dark energy changes the game. Since it appears to accelerate expansion, it seems a 'Big Chill' is almost guaranteed. However, there is also evidence that dark energy has grown more powerful over time. If this is indeed the case, then expansion might continue to accelerate, overwhelming gravitation on ever smaller scales. This could lead to a 'Big Rip' in which galaxies, solar systems and eventually individual atoms are torn apart by the expansion of the space in which they exist. Fortunately, most measurements think such events are a long way off. Our universe, with all its wonders, will be around for a long time yet.

Giles Sparrow
Space science writer

Giles has degrees in astronomy and science communication and has written many books and articles on all aspects of the universe.

Large clusters like Abell 2744 are the biggest objects governed entirely by gravity

The Big Bang theory

It's our best model of how the universe works, but where did it come from?

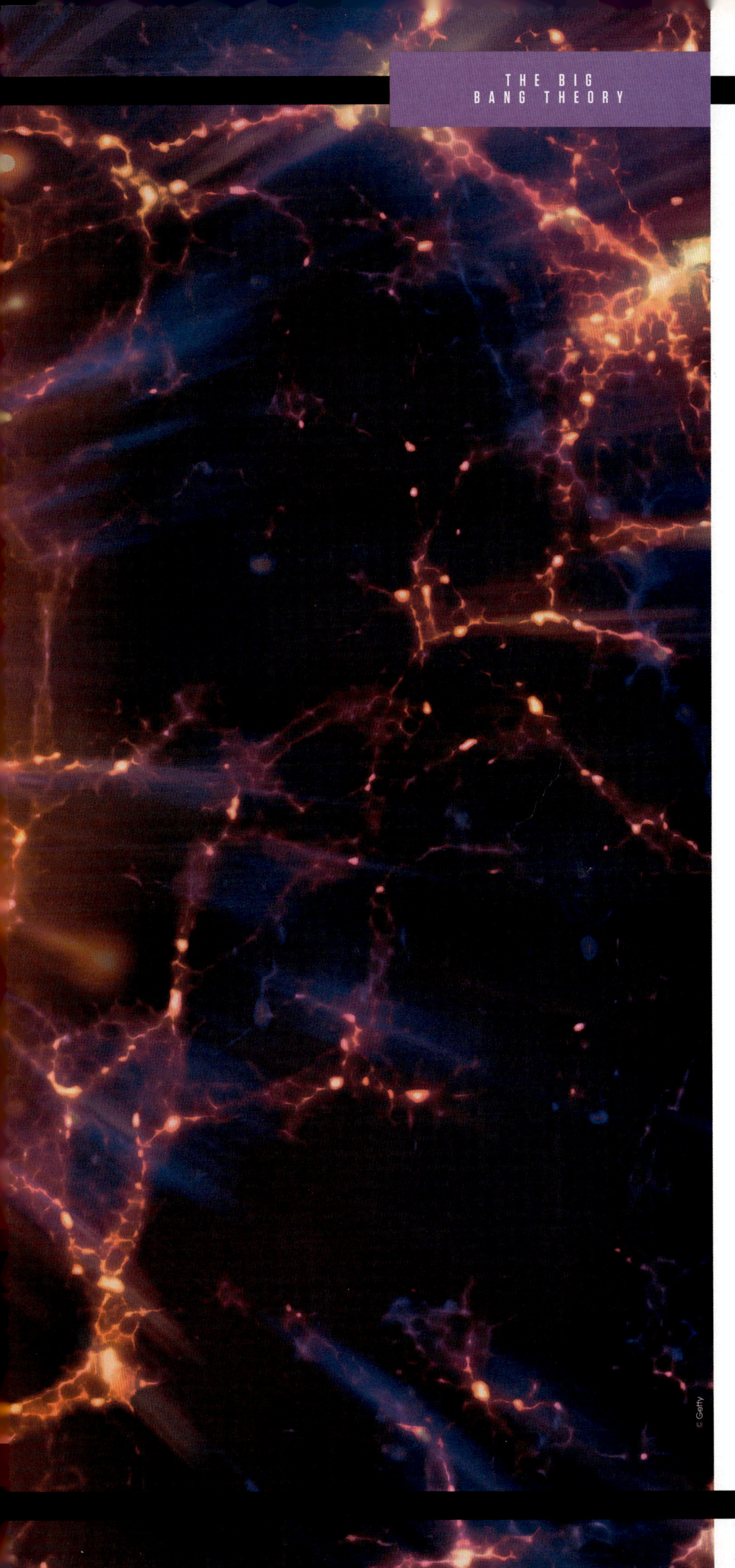
© Getty

People have always sought out 'big-picture' theories that explain how the universe began, what it looks like on the largest scales and how it evolves over time. In the past, such theories were often based more on human imagination than anything else. But our present best contender, the Big Bang theory, is much better than that. It's based on a mixture of observational evidence and a mathematical understanding of how space and matter behave on very large scales, and most astronomers believe it probably comes quite close to the truth.

There are two reasons we no longer need to rely on pure imagination to visualise the evolution of the universe. First there's the fact that we can actually see into the distant past. That's because light travels at a finite speed, so when a telescope shows us a galaxy a billion light years away, we're seeing it as it was a billion years ago. The second important factor is the universality of the laws of physics. This means we can study physics in laboratories here on Earth and know that exactly the same principles must apply to the rest of the universe as well.

Putting state-of-the-art observations and some very sophisticated physics together is what's given us the Big Bang theory. According to this, the universe began approximately 13.8 billion years ago as an infinitesimally tiny point, smaller than the smallest subatomic particle, with an unimaginably high density and temperature. From this minuscule beginning the universe rapidly expanded in size, eventually forming all the stars and galaxies we see today.

For many, there's something very appealing about the Big Bang theory, quite apart from its scientific veracity. With its image of a single, dramatic moment of creation, it conforms to earlier mythological and religious accounts, which gives it an air of familiarity for non-scientists. On the other hand, in its early days the theory met with opposition from some members of the scientific community, who felt instinctively that the universe should be constant and unchanging. The eminent British astronomer Arthur Eddington, for example, wrote that "the notion of a beginning is repugnant to me... I simply do not believe that the present order of things started off with a bang." It was another British scientist, Fred Hoyle, who actually coined the term 'Big Bang' to describe a theory he disagreed with.

Observing the cosmic microwave background

We can't see the Big Bang itself, but the CMB is the next best thing

COBE
The first space mission designed to study the CMB was NASA's Cosmic Background Explorer, launched in 1989. It found the CMB is much more uniform across the sky than expected.

WMAP
A second NASA spacecraft, the Wilkinson Microwave Anisotropy Probe, mapped the CMB between 2001 and 2010, confirming the astonishing uniformity.

Planck
Planck was the European Space Agency's follow on to WMAP. Operating from 2009 to 2013, it remapped the CMB with higher sensitivity and resolution than its predecessor.

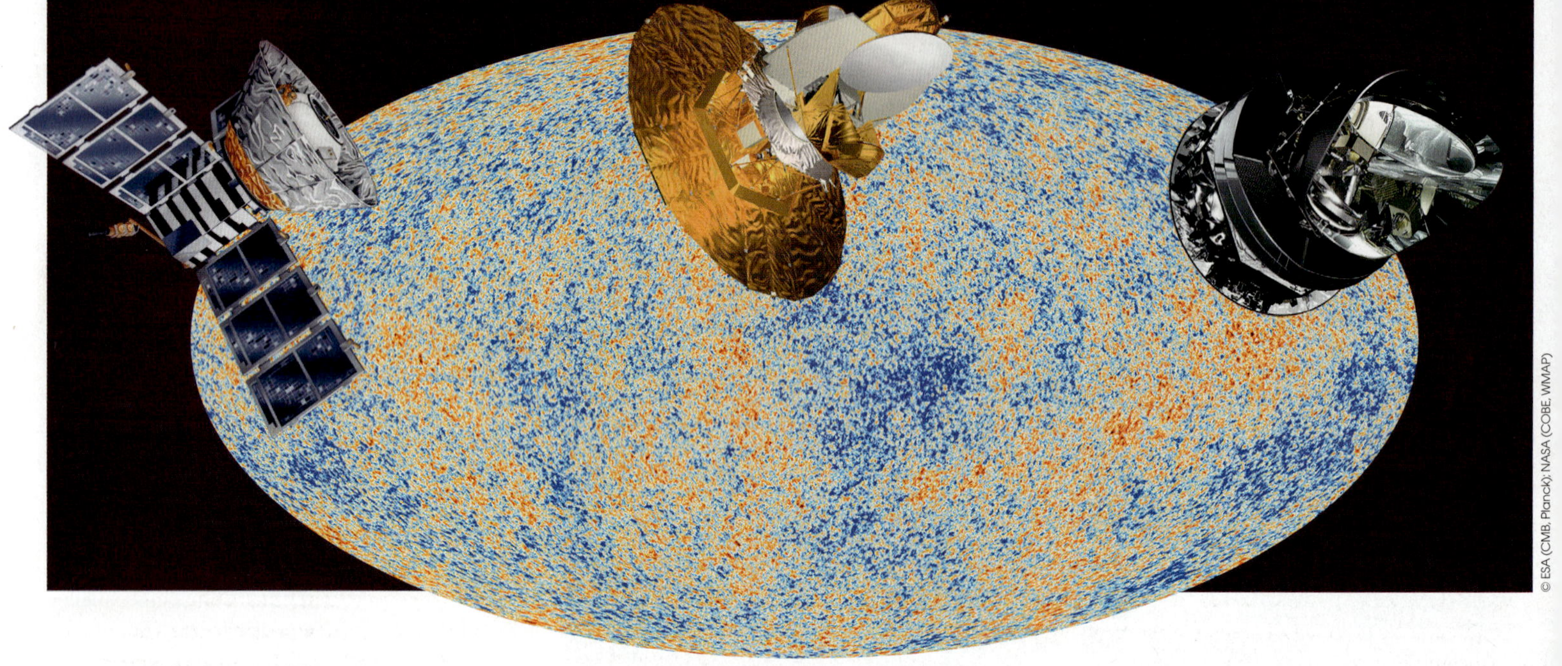

> **"The notion of a beginning is repugnant to me... I simply do not believe that the present order of things started off with a bang"**
> **Arthur Eddington**

Both Eddington and Hoyle were familiar with Einstein's theory of general relativity, which describes how space and matter behave on large scales. In its earliest form, the Big Bang theory originated in the 1920s as a solution of Einstein's equations – but it wasn't the only solution. The same equations also permit a perfectly static universe, so observational evidence was needed before scientists could choose between the options.

The first step in establishing the reality of the Big Bang was the discovery that the universe is expanding. This key breakthrough was made in 1929 by Edwin Hubble, who compared two sets of data for a number of galaxies beyond our own: their distances, based on the observed brightness of individual stars in them, and the speed at which they're moving away from us, based on spectroscopic measurements. To his astonishment, he found that the further away a galaxy is, the higher its recession speed. That could mean only one thing – that the whole of space is in a state of expansion. You can think of it like a fruit cake in the oven. As the dough rises, all the raisins move further apart from each other – but in this case these raisins are galaxies.

While an expanding universe is consistent with the Big Bang theory, it doesn't necessarily require it. As counter-intuitive as it sounds, you can have a universe that's in a constant state of expansion and yet eternal and unchanging at the same time. It just needs a small amount of new matter – a few hundred atoms per year per galaxy – to be continuously created in order to maintain a constant average density as the universe expands. This 'steady-state theory', championed by Hoyle and others, became a serious competitor to the Big Bang in the mid-20th century.

Yet the Big Bang prevailed in the end, and the steady-state picture fell by the wayside. Thanks to further observational evidence, we now know that Hoyle was wrong in assuming the universe has always looked the way it does today. Instead it's evolved over time, just as the Big Bang theory predicts. For example, Hubble images of the most distant galaxies, which we see not long after they were formed, look distinctly different than the mature galaxies we see closer to home – they're smaller and more irregular. However, that's a relatively recent discovery, and the real clincher came back in the 1960s with the discovery of the cosmic microwave background (CMB). This proved to be the kiss of death for the steady-state theory.

The CMB is sometimes described as the 'echo of the Big Bang', and if that were true then it would clearly be a great way to prove the theory correct. But the reality isn't that simple. For one thing, when Fred Hoyle put the 'bang' in Big Bang, he wasn't thinking of a literal sound, but a metaphor for the sudden onset of the universe's expansion from an initial, highly compressed state. There's no question of 'hearing' the bang, or an echo of it, but could a sufficiently powerful telescope see all the way back to that point, 13.8 billion years ago?

Unfortunately, even that isn't possible. When the universe was younger and smaller, it was also hotter, and in its earliest stages it was filled with a glowing, super-hot plasma like the Sun. That's something a telescope can never penetrate, because the plasma scatters light in the same way that clouds scatter sunlight. Just as we can't see above the cloud base, a telescope can't peer further back in time than the so-called 'surface of last scattering', around 380,000 years after the Big Bang. It's radiation from this surface – coming at us more or less uniformly from all directions – that forms the cosmic microwave background.

The CMB is an example of a phenomenon that was predicted theoretically before it was observed experimentally. Robert Dicke of Princeton University realised it would be the perfect way to verify the Big Bang theory and kill off its steady-state rival. He was busy devising an experiment to detect it when, in 1965, he heard about a problem Arno Penzias and Robert Wilson were having in an unrelated experiment at the nearby Bell Telephone Laboratories. Dicke realised that Penzias and Wilson had inadvertently detected the CMB, providing astronomers with the best evidence yet for the Big Bang.

Combining Hubble's imagery of distant galaxies with satellite measurements of the CMB, we now have pretty good observational data covering the entire history of the universe back to within 380,000 years or so of the Big Bang. But what happened before then? Although nothing can be observed directly prior to the surface of last scattering, astronomers were able to compare the CMB observations with physics theories and computer models and make a best guess. And when they did, the result wasn't anywhere near as simple as theoreticians expected.

The problem is that the CMB is simply too uniform across different parts of the sky. Scientists found it impossible to reconcile this with the idea that the universe has always been expanding at the relatively slow rate we observe today. Instead they had to assume a very brief period of cosmic inflation, during which the universe grew at a truly enormous rate.

Cosmologists have worked out the exact figures for this inflation, but they're almost impossible to express in everyday terms. NASA makes a brave attempt, describing the inflationary phase as lasting "less than a trillionth of a trillionth of a second", in which time the universe expanded "from a subatomic size to a golf-ball size" at a rate "far faster than the speed of light". Einstein supporters might be taken aback by that last statement, but in fact his prohibition against faster-than-light speeds only applies to the measured velocity of one object relative to another. In this case it's the fabric of space itself that's expanding faster than light, and that's perfectly legal within the framework of relativity.

It was only after the end of the inflationary period, when the universe had settled down to a more sedate expansion rate, that conditions became 'normal' enough for matter – in the form of subatomic particles – to appear. As the expansion continued, the initially super-hot universe gradually cooled to temperatures where

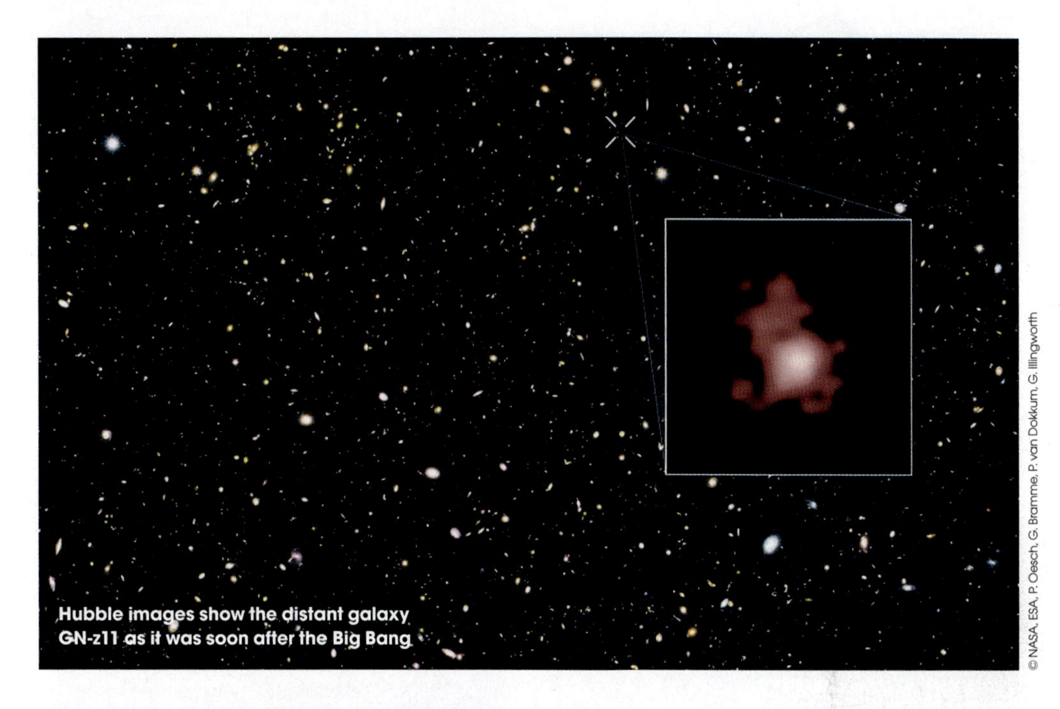

Hubble images show the distant galaxy GN-z11 as it was soon after the Big Bang

Expansion vs explosion

Although the Big Bang is often described as an explosion, that's a misleading image. In an explosion, fragments are flung out from a central point into a pre-existing space. If you were at the central point, you'd see all the fragments moving away from you with roughly the same speed. But the Big Bang wasn't like that. It was an expansion of space itself – a concept that comes out of Einstein's equations of general relativity but has no counterpart in the classical physics of everyday life. It means that all the distances in the universe are stretching out at the same rate. Any two galaxies separated by distance X are receding from each other at the same speed, while a galaxy at distance 2X recedes at twice that speed.

Webb and the Big Bang

A telescope is almost like a time machine, allowing us to peer back into the distant past. With the aid of Hubble, NASA has shown us galaxies as they were many billions of years ago. Hubble's successor, the James Webb Space Telescope, has the ability to look even deeper into the past. As of April 2023, it has been able to observe galaxies from around 13.4 billion years ago, less than 350 million years after the Big Bang. And unlike Hubble, which sees mainly in the visible waveband, Webb is an infrared telescope – a big advantage when looking at very distant galaxies. The expansion of the universe means that waves emitted from them are stretched out, so light that was emitted at visible wavelengths actually reaches us in the infrared.

these particles could combine into atoms, which eventually began to form stars and galaxies.

This brings us to the end of the 'Big Bang', insofar as the term refers to the birth of the universe, but the Big Bang theory goes beyond this, describing the whole evolutionary history of the universe to the present day and beyond. In this broader context, it turned out that observational astronomers had another surprise in store for the theoreticians.

In the original version of the Big Bang theory – the one that came from a solution of Einstein's general relativity equations back in the 1920s – the rate at which the universe expands isn't constant. It gradually slows down over time, pulled back by the combined gravity of all the matter in the universe. Turning this the other way around, it should be possible to calculate the total mass of the universe simply by measuring how quickly the expansion rate is slowing down. In the 1990s, a research team attempted to do exactly this by graphing expansion speed versus distance for a special class of astronomical object – Type 1a supernovae.

The result wasn't what they, or anyone else, expected. It turned out that the expansion of the universe isn't slowing down at all – it's speeding up. It was one of the most important cosmological discoveries of recent times, and it won them the Nobel Prize. It means that, on the largest scales, something is counteracting the effect of gravity, pushing galaxies apart faster and faster. As yet, no one knows what this mysterious 'something' is, but it's been given the spooky-sounding name of dark energy.

As weird as it sounds, dark energy doesn't contradict Einstein's theory of general relativity. It just means the particular solution that's been used so far isn't quite right. Einstein himself may have pointed the way to the correct solution, way before Edwin Hubble discovered that the universe is expanding. Einstein's own solution was a non-expanding static universe, but to achieve this he had to insert another factor – called the cosmological constant – into the equations. It was an idea that was abandoned after Hubble's discovery, with Einstein describing it as his biggest blunder. But a judicious choice of non-zero cosmological constant can produce an effect that, to all intents and purposes, is identical to dark energy – so Einstein may have been right after all.

Andrew May
Space science writer
Andrew holds a PhD in computational astrophysics and has written books on space and related subjects.

The grand design of nearby galaxies developed over time; closer to the Big Bang they're scrappier looking

Bubble universes?

Our Big Bang may have been just one of countless similar events

Everything in our universe originated after the Big Bang, so there are no conceivable observations we can make that would tell us anything about what happened before then. For some scientists this makes the question of what happened before the Big Bang meaningless, but for others it's no obstacle to theoretical speculation. One of the most intriguing suggestions is the concept of eternal inflation. The current Big Bang theory requires the universe to go through an initial period of very rapid inflation, which then transitioned abruptly to the more sedate expansion we see today. But what if when our own universe dropped out of this inflationary phase it was just a tiny bubble in a vast sea of inflating space? This is the idea behind eternal inflation, which would see a whole host of other bubble universes popping up at different times in other parts of the inflationary sea.

Artist's impression of the universe coming into existence

Dark energy: is it real?

Science writer Brian Clegg has doubts, as he explains

The discovery of dark energy has changed our whole understanding of how the universe expands and evolves over time – but does it really exist? One person who isn't so sure is Brian Clegg, author of *Dark Matter and Dark Energy*

What is the problem with the theory of dark energy?

I have a lot of sympathy with the idea that it doesn't exist at all. The original measurement was based on a correction of the data to handle the fact that both the Milky Way and supernovae are moving, but there are huge assumptions required to do this. Some recent work, using a much larger dataset, eliminated some of these assumptions and produced no dark energy effect.

What research could be done to resolve things?

The study I mentioned looked at 740 supernovae, compared with about 110 in the original study. We need to get more data still, and more accuracy as equipment is upgraded, to get a clearer picture.

What happened before the Big Bang?

Could there have been a time before the birth of the universe?

Could there have been a time before the Big Bang? In other words, could the universe have existed before it even began? May there have even been previous universes? Such ideas, once the preserve of high-concept science fiction and philosophical debates, are gaining a new scientific credibility in the 21st century. Some cosmologists are wondering if the Big Bang was merely an intermediate phase and not the true start of the universe at all. Theories such as the ekpyrotic universe, 'Big Bounce' models and cyclic cosmology have been around for a while, but data from sensitive space probes could put some of these on a firmer footing. But what exactly was the Big Bang, and why are some scientists now changing their minds about it?

The widely accepted standard cosmological model states that the universe came into being from a superhot, superdense state that was no bigger than an atom and made of pure energy. Not much about that is contentious, but things get precarious with what happens next. This object, known as the 'initial singularity', is thought to have been timeless and dimensionless; there was nothing 'outside' of the singularity to speak of. Then, 13.82 billion years ago – a figure obtained from NASA's Wilkinson Microwave Anisotropy Probe (WMAP) and European Space Agency's (ESA)Planck satellite – this microscopic singularity expanded rapidly to the size of a football. This was the 'Big Bang'. But it wasn't an explosion. The universe never exploded into being. Rather, this initial expansion from microscopic quantum fluctuations birthed space and time and seeded the large-scale structure of the universe. This 'Big Bang' model has served cosmology well for over 80 years, but there have always been unanswered questions.

Despite the Big Bang theory being the cornerstone of cosmology, a theory called cosmic inflation was proposed in the 1980s to address some of the problems with the original model, such as the horizon problem – how has the universe 'homogenised' on the largest scales when it hasn't existed for long enough to do so, given its enormous size? Cosmic inflation theory proposes an extremely rapid initial expansion rate of 10-32 seconds. The universe would then have continued expanding in line with the Big Bang theory.

Planck mapped the cosmic microwave background, relic radiation from the universe's creation

As the universe expanded, it also cooled, which resulted in energy condensing into matter known as subatomic particles. This transformation of energy into matter, predicted by Einstein's theory of special relativity, is described by the most famous equation in science: E=mc2. The universe, still seething and hot, was then a dense morass of quarks and electrons, with photons of electromagnetism, including those of visible light, trapped within it. After 380,000 years this still-expanding universe cooled enough for the first chemical elements, hydrogen, helium and lithium, to form. The quarks turned into the protons and neutrons of atomic nuclei, capturing free-travelling electrons in the process to make atoms. This was the point at which all of the trapped photons of the electromagnetic spectrum could travel unhindered. In other words, the universe became transparent. But it was still dark; it took another 400 million years for the first stars and galaxies to form.

Dense hydrogen and helium gas clumps collapsed under gravity, possibly collecting within a large 'dark matter halo', until atomic nuclei in their cores began fusing together, known as thermonuclear fusion, which released large amounts of energy as the first stars came alight. Galaxies formed within these haloes.

It's strange to think that our universe could have existed before any of these events, but the Big Bang wasn't always accepted. Eminent

What is a singularity?

Singularities are regions of space and time with extreme gravity, where not even light can escape, and infinite density. They are thought to exist inside black holes, and our universe is thought to have started from one, too. Although predicted by the general theory of relativity, neither that or quantum mechanics can explain singularities. They still remain truly mysterious to science.

What is the Big Bang?

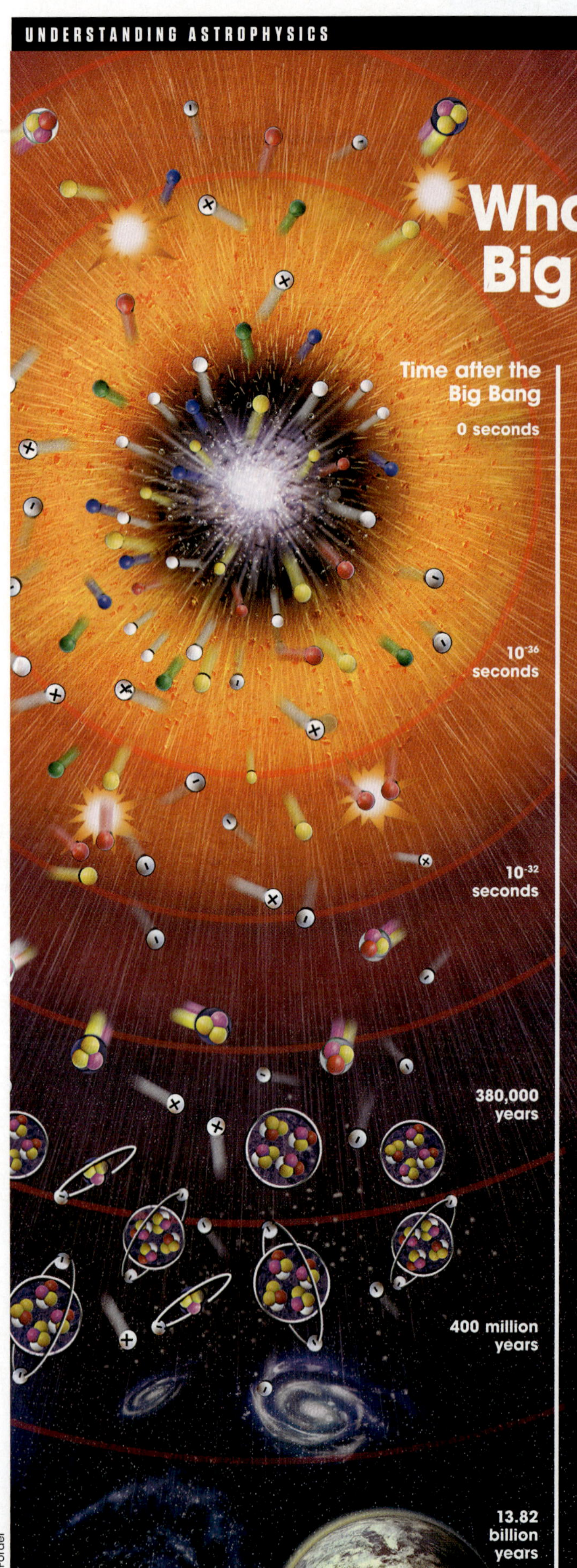

Time after the Big Bang
0 seconds
The beginning
The absolute beginning of our universe according to the Big Bang theory, which started out as a dense, hot, timeless, dimensionless singularity.
10^{-36} seconds
Cosmic inflation
A rapid expansion phase increased the size of the universe from that of an atom to a football. The universe was made of pure energy.
10^{-32} seconds
Cooling and quarks
After inflation ended, the universe cooled enough for subatomic quarks, electrons and other particles to form from the available energy.
380,000 years
Atoms form
Further expansion meant subatomic particles formed atoms. Hydrogen, helium and lithium filled the universe, which became transparent.
400 million years
First stars and galaxies are born
Gas clumps collapsed under gravity to form the first stars. They formed inside galaxies within dark matter halos.
13.82 billion years
Present day
Star formation and destruction created and spread chemical elements throughout space. That in turn created planets.

Millikan, Lemaitre and Einstein at Caltech in 1933

British astronomer Sir Fred Hoyle, who coined the term 'Big Bang' in a BBC radio interview in 1949, actually hated the idea. So why did it take such a hold in cosmology? In 1912, American astronomer Vesto Slipher saw that the spectra of galaxies were Doppler shifted towards the red end of the electromagnetic spectrum. This showed they were moving away from us at speed. Then, in the 1920s, Alexander Friedmann, a mathematician in the Soviet Union, and Belgian astronomer Georges Lemaître both independently proposed the idea of an expanding universe, which could explain Slipher's observations. But reception to the idea was lukewarm. Even Einstein – upon whose general theory of relativity their hypothesis was based – didn't accept the idea at first.

In 1929, Edwin Hubble showed that the recession speeds of galaxies actually increased with their distance from Earth. This meant that if the universe was a movie played backwards, all galaxies would have once 'existed' at the same point in space and time. Friedmann and Lemaître were vindicated, and the speed-distance relationship became known as 'Hubble's law'. In light of all this, English astronomer Arthur Eddington invited Lemaître to speak in London, calling his solution 'brilliant'. Lemaître posited the idea of a universe expanding from a single point, which he described as a 'primeval atom' or an 'exploding cosmic egg'. This is what cosmologists now call the initial singularity, the point of the Big Bang – although it wasn't actually an atom, or an egg.

Unlike Einstein and others, Hoyle actually had no problem with an expanding universe. What he hated was the idea of a 'beginning'. As an avowed atheist, Hoyle couldn't accept a point of creation, and thus a potential 'creator'. He clung doggedly to steady-state theory: the idea that the universe had always existed and was perpetually creating and destroying. But Hoyle was on the losing team. In 1948, American cosmologists Ralph Alpher and Robert Hermann predicted a background radiation to space – the residual heat 'echo' just before the universe became transparent 380,000 years after the Big Bang. As space had expanded for billions of years since, this radiation's wavelength should have been stretched into the microwave region.

Just 14 years later it was finally discovered by Arno Penzias and Robert Wilson using the Holmdel Horn Antenna. Initially believing it to be caused by bird droppings, they soon saw the spectrum of this cosmic microwave background (CMB) matched the predictions of the Big Bang model. Steady-state theory had no explanation for the CMB, and was therefore royally defeated. Alongside the work of Slipher, Friedmann, Lemaître and Hubble, Penzias and Wilson's evidence showed the universe had an origin after all and had been expanding. Big Bang theory was king.

And yet despite its enormous success, there's always been something that scientists have never liked about the Big Bang: it doesn't explain the initial singularity. Where did it come from? Why is it simply assumed to have been timeless, dimensionless and infinitely dense? Scientists hate assumptions, especially regarding the big questions. Even cosmic inflation theory, developed by physicists Alan Guth, Andrei Linde and Paul Steinhardt, which successfully ironed out some of the problems of the Big Bang, couldn't explain the singularity. As a result, alternatives to these cosmological cornerstones have been proposed, and it's from this that the idea of a pre-Big Bang existence has arisen. Strangely, these ideas may even be supported by the very same CMB data that supports the Big Bang theory.

In 2001, Steinhardt worked with Neil Turok, Justin Khoury and Burt Ovrut on the 'ekpyrotic' model of the universe – an alternative to inflation. In their original hypothesis, the universe was birthed from a collision between two multidimensional membranes, or 'branes', floating through a higher dimension of space. After the universe was created from the collision, the ekpyrotic phase would occur. This would also apply to a contracting brane. Imagine a prolonged contraction of a previous universe eventually collapsing back into a singularity before restarting again as our present universe in a typical 'Big Bang' scenario. The conditions for our universe – its fundamental laws and seeds for a future large-scale structure – would have been set in the previous universe, and not by inflation. This scenario seems quite exotic, but more up-to-date forms of the ekpyrotic model mostly do away with these multidimensional branes and other exotica. The newer models simply apply the physical constraints of the Big Bang theory.

A researcher working with one such form of the ekpyrotic scenario is Dr Yi-Fu Cai of McGill University in Canada. He says: "Since Neil, Paul et al proposed their original scenario, the physical picture is very clear. Their cosmological model is able to dilute unwanted relics [of the Big Bang] via the 'ekpyrotic phase'. But the universe is still expected to pass directly through the singularity from the contracting to expanding phases without that being removed." What he means is that in the multidimensional scenario, although ekpyrosis can 'smooth out' some problems like cosmic inflation can, the

Once atoms formed in the early universe, the resulting gas clouds collapsed to form stars

Supermassive black holes may be gateways to other universes

singularity is still present and the physics surrounding that are as problematic as ever. But Cai's work, performed with Professor Robert Brandenberger, head of McGill's high-energy theory group, does away with singularities entirely.

In their model, a previous universe collapsed until it could go no further and then 'bounced' out as a new universe. "In our scenario, the whole of cosmic evolution then becomes smooth. The physics around the bounce, including the CMB and perturbation, are well-controlled and calculable," he says. By removing the singularity, a lot of associated problems are also removed. Cai and Brandenberger's work also predicts the existence of the CMB and the microscopic perturbations that grow to become the universe's large-scale structure. This is even consistent with data from the WMAP and Planck probes. Could the CMB contain hints of a previous universe that we could detect?

Many cosmologists have asked that very question. One scenario that has an answer is a cyclic model of cosmology. With their concept of universes perpetually oscillating between expanding and contracting phases, cyclic cosmological theories have a lot in common with 'Big Bounce' models. The idea has been around since at least the 1920s, and one variant was even proposed by Einstein in 1930 after accepting Hubble's observations supporting an expanding universe. In the cyclic scenario, after an expansion phase the universe would slow, then stall, and would then contract back due to the gravitation of all the matter within it. This would culminate in a 'Big Crunch' or a 'Big Bounce', which would then be followed by a new expansion phase and so on. Einstein thought that this cyclical scenario could be a better, more long-term alternative to the simpler idea of an expanding universe with a single origin point.

But in 1934 American physicist Richard Tolman showed that such cyclic models couldn't work the way people wanted them to because of the second law of thermodynamics. Over time, the amount of entropy in an oscillating universe would only increase, and the amount of usable energy within it would only decrease. Every expansion would be slower and larger than the previous one, as each contraction phase would

only go back so far and less energy would be available for each new expansion phase. Conversely, previous phases would have started out smaller and smaller until you eventually returned to a 'Big Bang' scenario again anyway. But in the age of WMAP and Planck, the idea of cyclic, oscillating universes has re-emerged. One example, conformal cyclic cosmology (CCC), was developed by English physicist Sir Roger Penrose and Armenian mathematician Vahe Gurzadyan in 2010 and is based on the theory of general relativity. Using data from both WMAP and Planck and also the BOOMERanG (Balloon Observations Of Millimetric Extragalactic RAdiation aNd Geophysics) experiment, Penrose and Gurzadyan published results that purported to show extremely

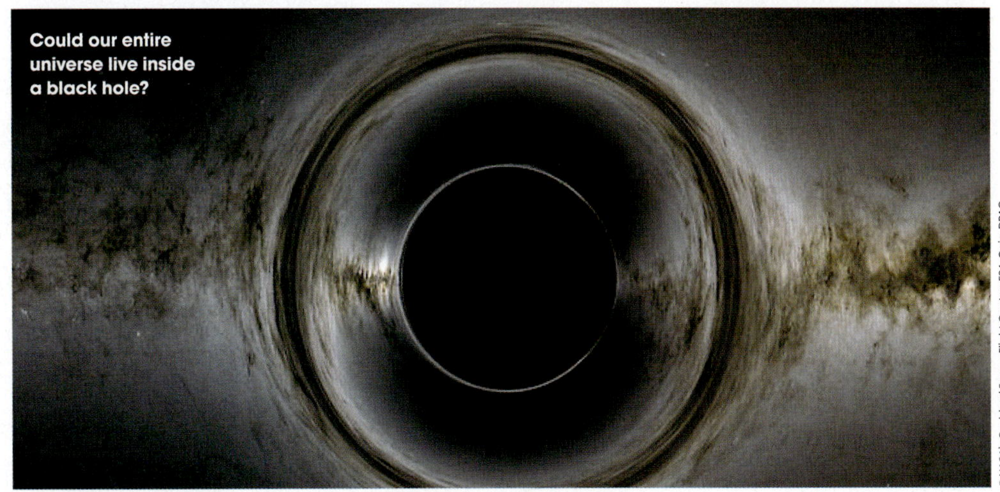

Could our entire universe live inside a black hole?

© NASA's Goddard Space Flight Center; ESA, Gaia, DPAC

"Our universe should obey the same laws as the universe in which the black hole exists"
Nikodem Popławski

faint concentric rings from previous cosmic cycles in CMB fluctuations – similar to ripples spreading out when you throw a stone into a pond. According to CCC, what we think of as the universe, the region that we can observe, anyway, is simply an 'aeon' or domain within an infinitely larger space-time.

Eventually, far into the future, once all the stars and galaxies have died out, all matter has dispersed and the supermassive black holes that lay at the centres of galaxies have evaporated, our aeon will have become completely smooth. But it will continue to expand and birth a new, larger scale aeon. CCC theory, which unlike previous cyclic models has no contraction phases, states that what we think of as inflation is simply the accelerating expansion of a previous aeon. But other cosmologists looking for concentric rings in the CMB haven't found anything significant yet. This may be because they used standard simulations to check against, whereas Penrose and Gurzadyan adopted a nonstandard approach.

Cyclic models like CCC remain controversial, and the Big Bang theory itself still has support. Physicist Lawrence Krauss wrote in 2012 that "the Big Bang picture is too firmly grounded in data from every area to be proven invalid in its general features". So could it be the true picture of the universe after all? Nikodem Popławski, a physicist at the University of New Haven, Connecticut, has developed a theory stating our universe originated from a black hole. Such extraordinary hypotheses have been a topic of speculation for years. "Our universe should obey the same laws as the parent universe in which the black hole exists," he says, adding that it should be possible to determine the size of the parent black hole by measuring temperature fluctuations in the CMB. "I published a paper that shows consistency with Planck's observations of the CMB. It also shows they aren't too sensitive to the black hole's initial size," he says.

And Popławski is now working on finding evidence for a black hole origin scenario. "If our universe was formed by a 'Big Bounce' in a black hole, its early expansion has specific dynamics that can be tested by measuring temperature fluctuations in the CMB from all directions in the sky," he says. Intriguingly, predictions that Popławski and his colleague Shantanu Desai of Garching, Germany, made of CMB fluctuations are consistent with the latest Planck data. And black holes rotate, so if our universe really was birthed from one, Popławski expects to see those effects, too. Echoing late cosmologist Stephen Hawking, he says: "Our universe could be the interior of a black hole existing in another universe. Black holes forming from stars and galaxies in such a universe create new universes. And so a universe can parent billions of baby universes, which are formed through black holes."

Kulvinder Singh Chadha
Space science writer

Kulvinder is a freelance science writer, outreach worker and former assistant editor of *Astronomy Now*. He holds a degree in astrophysics.

What caused the birth of the universe?

Many scenarios illustrate how our universe may have come into being

1 There was nothing

Strictly speaking, no scientist believes that our universe started from literally nothing. There always has to be something to cause another action to occur. However, many cosmologists are not yet convinced that there was anything before the Big Bang. Lawrence Krauss is one such scientist, and he's developed a theory of 'quantum nothingness' from which the universe could have originated.

The theory involves so-called 'virtual particles' that flit in and out of existence for fractions of a second in empty space. They are predicted by quantum theory and their effects can be observed. Krauss says that if you remove virtual particles from a region of space it still has an energy density – but it shouldn't. It's from this form of 'nothing' that our universe could have started.

BACKED BY
Lawrence Krauss
Origins Project Foundation

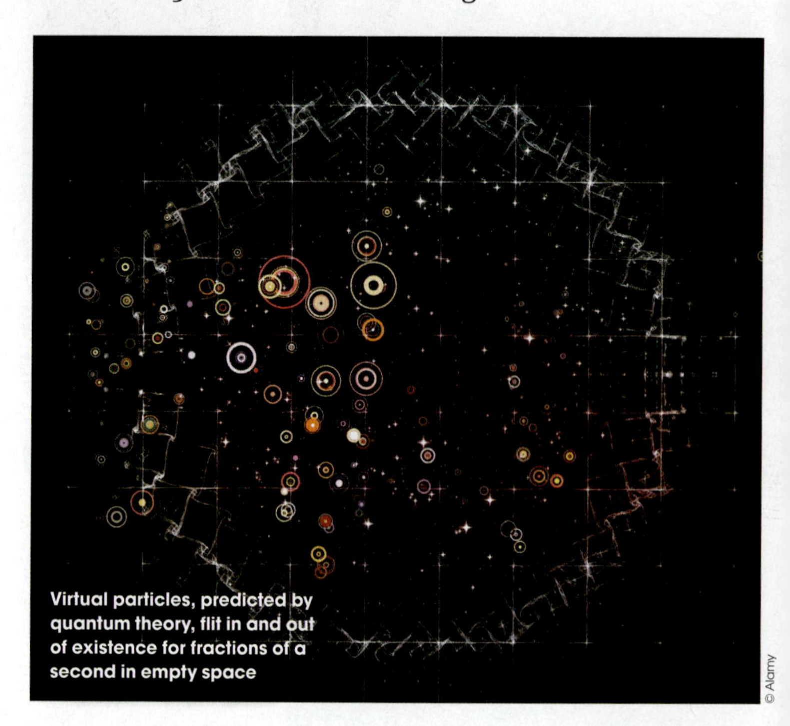

Virtual particles, predicted by quantum theory, flit in and out of existence for fractions of a second in empty space

2 There was another universe before ours

1 Big Bang and expansion
In the cyclic model, a universe may have a true origin as normal and go through an inflation and expansion phase.

2 Everything now stalls
In a cyclic universe the cosmos may expand to a point and then stall. This may happen from the gravitational effect of the matter within it.

3 Universe in reverse
Gravity takes over and the universe now contracts, with galaxies moving towards one another instead of further away.

4 Big Bounce and the next phase
Inevitably, the universe can only contract so far. A 'Big Bounce' initiates the next expansion phase. The process continues, with each phase getting larger and slower.

BACKED BY
Sir Roger Penrose
Oxford University and Wadham College

3 It's always been there

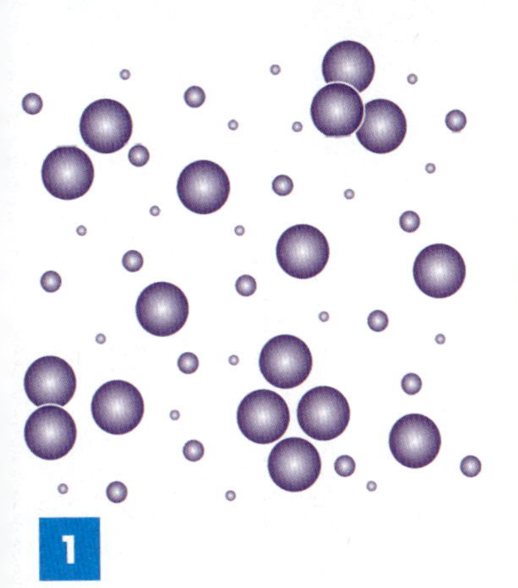

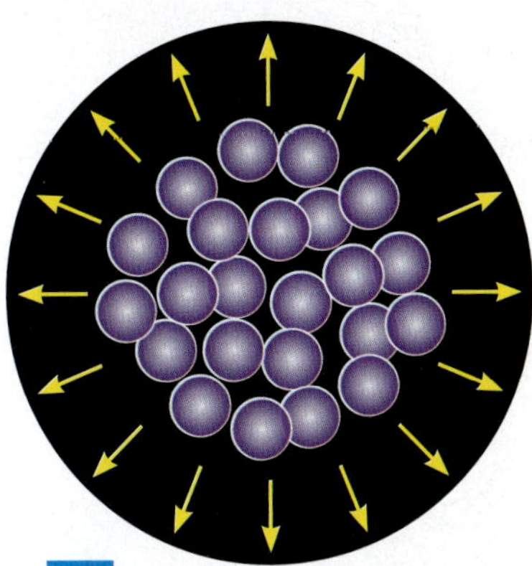

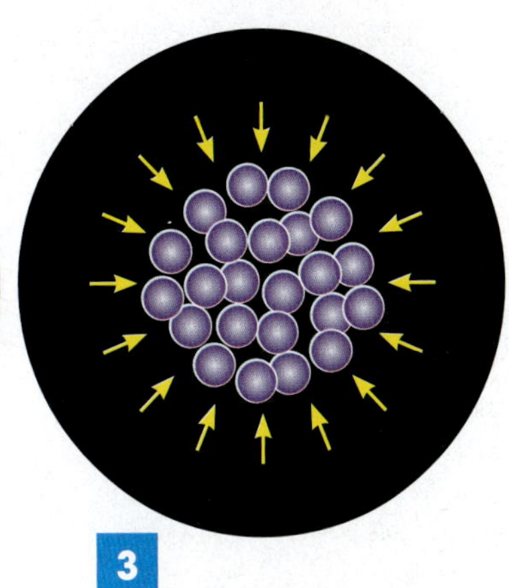

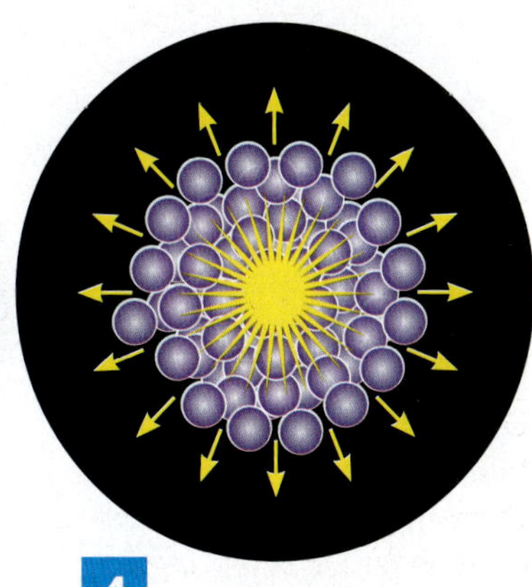

1 An empty, infinite universe
In this scenario, the universe has existed forever and was nearly empty for that time. Then gravity took over and matter started to clump together.

2 Expanding internal region
The density of matter is such in some regions that it forms an incredibly massive black hole, the internal region of which experiences expansion.

3 A Big Crunch
Inside the huge black hole, matter again collapses under the intense gravity and increases in density up to a limit imposed by physics.

4 The Big Bang
When it can't stand any more, quantum fluctuations cause the matter to expand outwards in a typical Big Bang scenario within its black hole universe.

BACKED BY
Gabriele Veneziano
CERN and Collège de France

4 It's one universe of many

Eternal inflation theory was proposed in 1983 by physicist Paul Steinhardt as an extension of the cosmic inflation and Big Bang theories. Alan Guth, Andrei Linde and Steinhardt originally developed cosmic inflation theory to explain some problems with the Big Bang model, and it involved an exponential but rapid expansion of our universe.

The cause of cosmic inflation still remains somewhat vague, but for eternal inflation, Guth proposed in 2007 the existence of a 'false vacuum' or region of space with a positive energy density – similar to expanding bubbles forming in a boiling liquid. In this manner, certain regions of space-time, or 'universes', would be affected by their own form of cosmic inflation before the positive vacuum moved on to another region. As of yet, this scenario lacks evidence, but if true then our universe could exist as a nodule on another universe as part of a 'multiverse'.

BACKED BY
Professor Alan Guth
Massachusetts Institute of Technology

Another theory suggests that our universe is one of many that exist parallel to one another and is part of a multiverse

What is a light year?

How we measure vast distances across the universe

A light year is a measurement of distance and not time, as the name might suggest. A light year is the distance a beam of light travels in a single Earth year, or around 9.7 trillion kilometres (6 trillion miles). On the scale of the universe, measuring distances in kilometres or miles just doesn't cut it. In the same way that you may measure the distance to the supermarket in the time it takes to drive there, astronomers measure the distances of stars in the time it takes for their light to travel to us. For example, the nearest star to our Sun, Proxima Centauri, is 4.2 light years away.

Unlike the speed of your car when travelling on different roads or in traffic, the speed of light is constant throughout the universe and is known to high precision. In a vacuum, light travels at 1,079,252,849 kilometres (670,616,629 miles) per hour. To find the distance of a light year, you multiply this speed by the number of hours in a year – 8,766. Therefore one light year equals 9.5 trillion kilometres (5.88 trillion miles). At first glance this may seem like an extreme distance, but the enormous scale of the universe dwarfs this length.

Measuring distances in kilometres or miles at an astronomical scale would be extremely cumbersome and impractical. Starting in our cosmic neighbourhood, the closest star-forming region to us, the Orion Nebula, is a short 12,651,053,184,000,000 kilometres

"When you observe something a light year away, you see it as it appeared exactly one year ago"

(7,861,000,000,000,000 miles) away. Or, more simply, 1,300 light years away. The centre of our galaxy is about 27,000 light years away. The nearest spiral galaxy to ours, the Andromeda Galaxy, is 2.5 million light years away. Some of the most distant galaxies we can see are billions of light years from us.

Measuring in light years also allows astronomers to determine how far back in time they are viewing. Because light takes time to travel to our eyes, everything we view in the night sky has already happened. In other words, when you observe something a light year away, you see it as it appeared exactly one year ago. We see the Andromeda Galaxy as it appeared 2.5 million years ago. The most distant object we can detect, the cosmic microwave background, is also our oldest view of the universe, occurring just after the Big Bang some 13.8 billion years ago.

Astronomers also use parsecs as an alternative to light years. Short for parallax second, a parsec comes from the use of triangulation to determine the distances of stars. To be more specific, it's the distance to a star whose apparent position shifts by one arcsecond in the sky after Earth orbits halfway around the Sun. One parsec is equal to 3.26 light years.

The light year can also be broken down into smaller units of light hours, light minutes or light seconds. For instance, the Sun is just over eight light minutes from Earth, while the Moon is just over a light second away. Scientists use these terms when talking about communications with deep-space satellites or rovers. Because of the finite speed of light, it can take more than 20 minutes to send a signal to the Perseverance rover on Mars, and another 20 to get one back. Whether it's light years or parsecs, astronomers will continue to use both to measure distances in our expansive and grand universe.

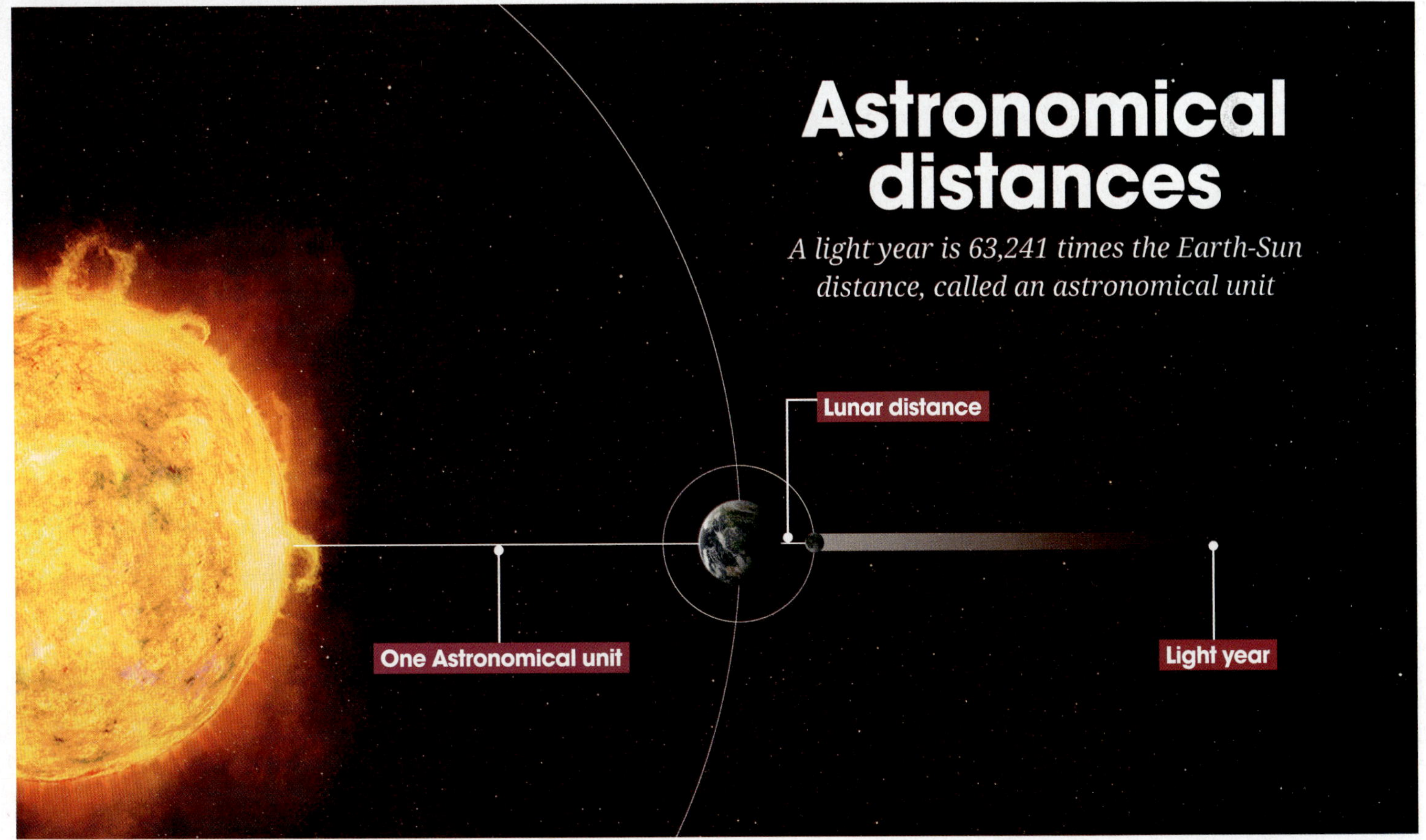

How to build a galaxy

The making of these billion-star structures has been puzzling astronomers for decades. All About Space puts together the pieces for building a galaxy

The universe is packed with galaxies. Everywhere we look we see galaxies crammed into the cosmos, grouped in clusters and great sheets that are littered throughout space-time. The most distant galaxies ever seen have been identified by the Hubble Space Telescope as being 13.4 billion light years away. Yet these are not even the most distant galactic structures out there. It's the first galaxies that hold the record for being the furthest away from us. However, we're yet to see them. To do so we will need the infrared prowess of the James Webb Space Telescope (JWST), which has the ability to see galaxies as they formed just 300 or 400 million years after the Big Bang – the event that threw the cosmos into existence.

Understanding how galaxies form is a bit like trying to put a jigsaw puzzle together. Think of each galaxy as being a piece of that puzzle. Because galaxies are so old and evolve so slowly, when we see a galaxy in the night sky we are just seeing a single snapshot of its long life. However, the galaxies are all at different stages of their evolution, so if we can put all these snapshots together – just like arranging the pieces of a puzzle – we can build an overall picture of how galaxies like our

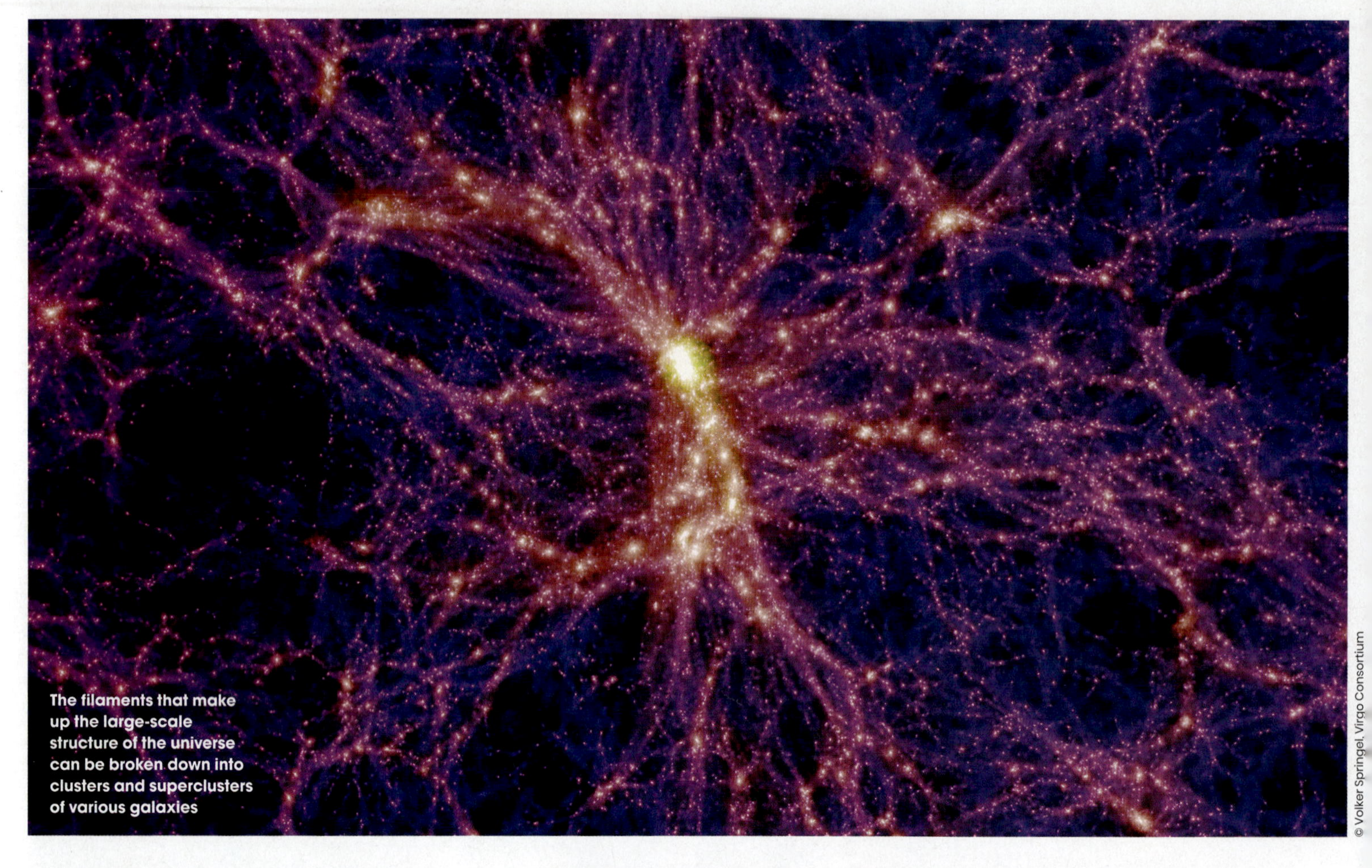

The filaments that make up the large-scale structure of the universe can be broken down into clusters and superclusters of various galaxies

own Milky Way grew into the star, dust, gas and dark matter-packed structures we see today.

We've actually only known that there are galaxies in the universe beyond our own Milky Way for around a hundred years. Before that time astronomers thought that the weird objects they dubbed 'spiral nebulae' were still part of our galaxy. Their telescopes weren't powerful enough to resolve individual stars in these objects, although when astronomers looked at the light coming from them, they had all the evidence they needed to confirm that these blobs in the night sky were made up of many stars. However, in 1912, American astronomer Vesto Slipher found that the light being thrown out by the spirals was Doppler shifted towards redder wavelengths, showing that the 'spiral nebulae' are moving away from us and take on a red tint. Doppler shift is the compression or stretching of light waves as an object moves towards or away from us. You might not have realised it, but you have experienced a Doppler shift before since it also happens with sound waves. When a police car or ambulance has raced past you with its siren blaring, the pitch of the sound it makes changes depending on its distance. At first it is higher, becoming lower as it moves away since the sound waves become compressed and then stretched.

"We've actually only known that there are galaxies in the universe beyond our own Milky Way for around a hundred years"

In 1925, Edwin Hubble announced he had discovered that the spiral nebulae were all galaxies, or 'island universes', far beyond our own. He achieved this thanks to the biggest telescope in the world at the time – the 2.5-metre (8.2-foot) Hooker Telescope at Mount Wilson Observatory in California. Hubble was able to resolve individual stars, including a specific type called a Cepheid variable. This type of star throws out light which varies according to its true brightness, and from this observation astronomers can work out their distance. It was the Cepheids found in our neighbouring galaxy of Andromeda that allowed astronomers to measure the distance to our closest spiral as 2.5 million light years away. Coupled with Slipher's discovery that galaxies are all moving away from us, scientists quickly realised that once upon a time they must have been much closer together than previously anticipated.

Hubble's next step was to classify all of the galaxies in an effort to understand them as much as he could. It was from this that he created his famous tuning fork diagram. He built this diagram so that its handle is made up of elliptical galaxies, which are galactic structures with ovoid shapes. Hubble referred to these as early-type galaxies because he believed that all galaxies began their lives as ellipticals before evolving into one of two types of spiral galaxy, forming the prongs of his tuning fork. One type are regular grand-design galaxies with their graceful spiral arms curving away from a small central bulge, while the second type are barred spirals, whose spiral arms are connected with a long, straight bar running through their glowing centres. In fact, today it's believed that our very own galaxy has a bar running through its centre. Hubble thought that the elliptical galaxies were the bulges of spiral galaxies without the arms, which he assumed grew later. However, astronomers changed their minds about this after studying galaxies in more detail throughout the 20th century. They found that elliptical galaxies form when two or more spiral galaxies collide and merge, and it's spiral galaxies that are really the early ones.

So how do spirals form? The main ingredients are gas – predominantly hydrogen – and dark matter. Nobody knows what dark matter is, but we know it comes in the shape of giant blobs scattered throughout the universe. Some of these dark matter blobs are large enough to hold clusters of thousands of galaxies. The dark matter came first, forming these blobs, or halos, very soon after the Big Bang. The gravity of these halos began to attract hydrogen gas towards them, which began to flow like rivers along gravitational inclines created by the influence of dark matter into the cores of the blobs. There the hydrogen formed enormous spinning clouds, and the hydrogen and dark matter formed the embryo of a galaxy – think of the white and the yolk of an egg as the dark matter with the hydrogen at the core.

Because the dark matter and hydrogen mixture was spinning so fast, it flattened into a pancake shape, taking on the characteristics of a spiral galaxy's flat disc. Meanwhile, small pockets of hydrogen gas in the cloud collapsed to form the very first stars. These stars were gigantic – hundreds of times more massive than our relatively puny Sun – and they exploded very quickly as powerful supernovae. Stars are able to create elements in their cores, while the violence of exploding stars, known as supernovae, can form even heavier elements. When the first stars detonated, they spilled their guts into the young galaxy around them, enriching it with these heavy elements. Over time, enough of these elements would build up to form asteroids, moons and planets. When we look at galaxies today, including our own Milky Way, we see vast lanes of black dust. This dust is comprised of elements made inside the nuclear furnaces of stars, dating all the way back to the first stars that existed about 13.5 billion years ago.

Today we find that the oldest parts of spiral galaxies are their bulges. In these central regions, most of the gas has been used up and the stars that exist there are crammed together and more red than the combined light of the stars in the spiral arms, which are instead dominated by hot, young stars. The exception is in the few tens of light years immediately around the supermassive black hole that lies in the middle of every large galaxy, where the gas is dense enough to keep forming new star clusters made of massive stars.

The black holes in the centres of galaxies are enormous. The black hole that lies at the heart of the Milky Way is 4 million times more massive than the Sun. In other galaxies, black holes can be tens or even hundreds of millions of times more massive. The biggest galaxies of all – the giant ellipticals found in the centres of galaxy clusters – have central black holes with masses over a billion times that of the Sun, as is the case with the galaxy Messier 87 in the heart of the Virgo Cluster.

It's thought that galaxies are made from the collapse of protogalactic clouds of dense hydrogen and helium gas in the early universe

Everyone knows that black holes like to consume matter – that's how they grow so big. But black holes can't eat everything that is served their way, and sometimes they spit out their food. As gas flows towards a black hole, it whirls around into a disc of material spiralling into the centre. However, the gas brings magnetic fields with it that become wrapped up around the black hole by the swirling gas. Eventually the magnetic fields become so strong that they can actually begin to funnel away charged particles, atoms, protons, electrons and ions into jets that are so energised they race away from the black hole at almost the speed of light. We can even see one of these jets coming from the black hole in Messier 87.

The level of black hole activity can depend on many factors, such as the mass of the black hole and the amount of gas falling into it.

The stars in galaxies are made from the collapse of clouds of gas and dust under their own gravity

Galactic evolution

How their different sizes can affect how galaxies form

1 A lonely cloud of gas
In order for what astronomers call a small galaxy to be made, a relatively large and isolated gas cloud is needed.

2 The making of stars
Under gravity, the cloud will collapse because there's not enough pressure from the gas itself to fight against it. Baby stars are made in the fight between gravity and pressure.

3 Forming a disc
The matter spins quickly, causing a flattened disc-like structure. At the centre is a bulge where the older first-generation stars can be found. The rest of the disc is teeming with younger stars.

4 A galaxy with arms
Internal processes make the arms and bars found in spiral galaxies. However, if conditions are more favourable, a lenticular galaxy – an intermediate between an elliptical and a spiral – is made instead.

5 A team of gas clouds
Small clouds of gas collapse early on to form the galaxy's very first stars.

6 A party of stars
These gas clouds, with their newly formed stars, clump together to make a larger cloud with a party of stellar populations.

Small galaxies

1

2

3

Large galaxies

5

6

7

7 Gaseous add-ons

There isn't much spinning going on during the making of a large galaxy. Instead the merging of nearby gas clouds stops any chances of a disc-like structure forming.

8 A gigantic galaxy

Since most of the gas needed to make a new generation of baby stars was mopped up, no more can be made. What's left is a gigantic elliptical galaxy that's dominated by old stars.

Spiral galaxy NGC 7678 features a peculiarly prominent spiral arm

Our Milky Way's supermassive black hole, for instance, is very quiet, with hardly any gas falling into it. Other spiral galaxies have more activity in their centres, with some emitting strong radio waves. However, the most active black holes are called quasars. The closest to us is 600 million light years away, but the majority existed in the universe over 10 billion years ago.

Quasars are fed by gas in two different ways: one is simply clouds of intergalactic gas falling onto a black hole in the centre of a galaxy. These clouds are clumps of gas and dark matter left over from the process of building galaxies over 13 billion years ago. The other way that quasars light up is more exciting – when two galaxies come hurtling towards each other and collide, it causes huge clouds of interstellar gas and stars to fall into their black holes. Sometimes the collision is a hit and run. The gravitational forces of each galaxy tear stars and gas out into long streamers that astronomers call tidal streams. These streams can sometimes be many hundreds of thousands of light years long.

When galaxies collide, it changes the future of the structures involved. Going back in time by 13 billion years, the Hubble Space Telescope is able to see the first galaxies growing by consuming smaller galaxies. This galactic cannibalism continues even today, although at a much slower rate. Even the Milky Way is eating smaller galaxies at this very moment, but they're not going near our black hole, so the centre of our galaxy isn't active. For example, the Canis Major Dwarf Galaxy is only 25,000 light years from Earth and is merging with our galaxy. It contains

Lighthouse of the cosmos

Quasars blaze radiation that can be seen from the other side of space

1 The 'calm' black hole
If a black hole is calmly sitting at the centre of its galaxy, it generally distorts the fabric of the universe around it. It leaves a dent in this sheet of space-time from which nothing – not even light – can escape.

2 A swirling disc of dust and gas
An accretion disc made of gas and dust circles the black hole. If the black hole isn't particularly active, then the matter won't fall into it.

3 Heating up
When the material falls into the black hole and reaches the event horizon – the point of no return – a lot of friction is created, superheating atoms and tearing them apart.

Galactic arithmetic

The basic space formulae for creating galaxies

© Tobias Roetsch

few stars because the gravity of the Milky Way has stripped most of them away.

Astronomers call such collisions minor mergers. The end result is that the smaller galaxy is swallowed up, increasing the mass of the larger galaxy. At the other end of the scale are the major mergers between two large galaxies of around the same size. When two spiral galaxies collide like this, it destroys their spiral structures and they merge into a giant elliptical galaxy – the opposite of what Hubble's tuning fork suggests. Amazingly, during a galaxy collision stars smashing together is extremely rare. The space between the stars is so large that the chances of two coming within each other's gravitational sphere of influence is very small. When our Milky Way undergoes the next phase of its growth and merges with the Andromeda Galaxy in 4 or 5 billion years, our Sun likely won't collide with another star from Andromeda. What will collide will be the huge clouds of interstellar gas that inhabit the spiral arms of both galaxies, igniting in a huge burst of star formation. We call this a starburst,

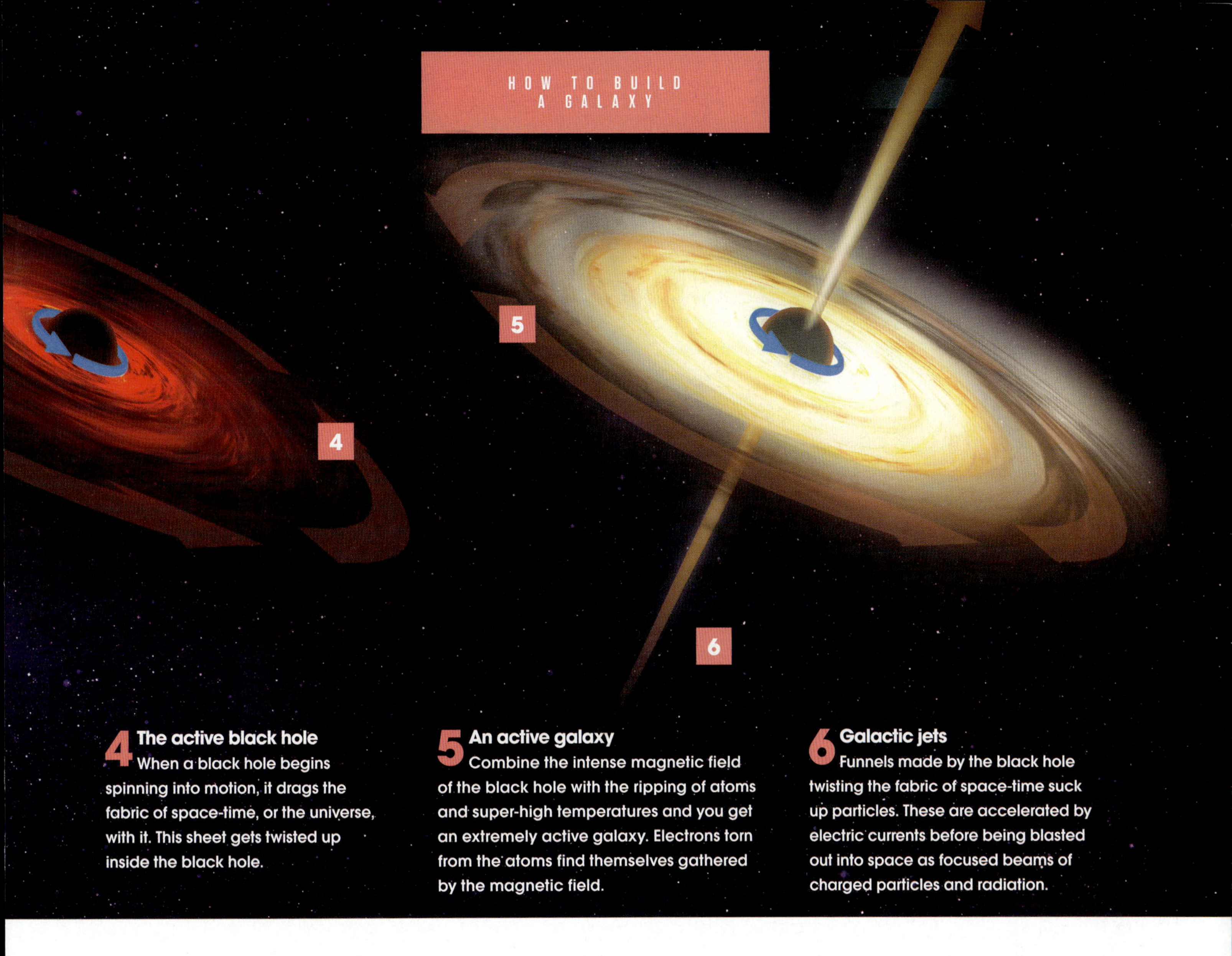

4 The active black hole
When a black hole begins spinning into motion, it drags the fabric of space-time, or the universe, with it. This sheet gets twisted up inside the black hole.

5 An active galaxy
Combine the intense magnetic field of the black hole with the ripping of atoms and super-high temperatures and you get an extremely active galaxy. Electrons torn from the atoms find themselves gathered by the magnetic field.

6 Galactic jets
Funnels made by the black hole twisting the fabric of space-time suck up particles. These are accelerated by electric currents before being blasted out into space as focused beams of charged particles and radiation.

and it can use up all the gas in a galaxy. This is why most elliptical galaxies, which form from mergers, have no star-forming gas left and haven't made any new stars in a very long time. All their short-lived, young stars exploded long ago, leaving ellipticals dusty and red with older, cooler stars.

No galaxies are being made today. All that galactic construction happened over 13 billion years ago, and ever since it has been a case of galactic evolution rather than galactic formation. There are still some crucial pieces of the jigsaw missing, such as whether supermassive black holes formed before the galaxies that exist around them or vice versa, why the disc turns into spiral arms and why these arms do not wind up as they rotate around the centre of the galaxy. Some scientists think that the spiral arms are not actually rigid appendages, but density waves where stars and gas are bunching up. This is a bit like a traffic jam on a motorway, and as soon as some stars hit the brakes and slow down, all the other stars and gas clouds bunch up behind them.

Although some of the pieces are missing, the jigsaw of how galaxies are made, grow and evolve is becoming clearer. We might not be able to see everything, but we can see enough to understand when and where galaxies came from.

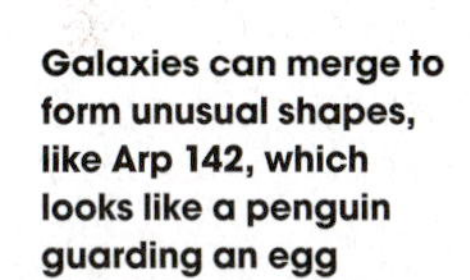

Galaxies can merge to form unusual shapes, like Arp 142, which looks like a penguin guarding an egg

Secrets of black holes

Places where the laws of physics are pushed to the extreme

Black holes are the most mysterious objects in the universe. They're places where physics is pushed to its most extreme, where light cannot escape and where space-time itself is twisted and even punctured, leading to the most incredible and counter-intuitive phenomena. A black hole is a region of space where gravity is so strong that nothing, not even light, can escape from its grasp. Within a certain proximity of one, closer than the black hole's 'event horizon', you'd have to travel faster than light to get away from it. Since nothing can go faster than light – at least as far as scientists know – then whatever falls down a black hole stays down the black hole.

The discovery of black holes dates back to Einstein's general theory of relativity. Einstein himself didn't predict the existence of black holes per se, but general relativity, which describes mass, space, time and gravity, provides the mathematical foundations for understanding them. These were realised by Einstein's German compatriot Karl Schwarzschild, who solved Einstein's equations to describe the gravitational field around a non-rotating, spherical mass and to determine the Schwarzschild radius, which is the size of a black hole's event horizon. In the 1960s, Roy Kerr solved Einstein's equations for a more realistic scenario – that of a black hole that's spinning.

We've already mentioned that light cannot escape a black hole, and the event horizon is its ultimate boundary of no return. Once something has crossed the event horizon's invisible boundary, it can never return from the black hole. Let's picture what is going on using an oft-mentioned analogy, that of a rubber sheet, which we have to imagine as being the fabric of space for this analogy to work. If you want to try this at home, a bedsheet held tight at each corner should suffice. Place a marble onto the sheet. In our example, that's Earth. Notice how it causes the sheet to dip a little. In general relativity, that dip in the fabric of space is called a gravitational well – it represents Earth's gravitational field. Now put a tennis ball on the sheet, imagining that it's the Sun. You'll notice that it creates a bigger dip than the 'Earth', not necessarily because it's larger, but because it has more mass. If you were to zip ball bearings past both the marble and the tennis ball, they'd need more energy to get past the tennis ball without falling into its steeper gravitational well.

Now, put a cannonball on the rubber sheet – if you're trying this at home you probably don't have a cannonball to hand, but see if you can use something suitably heavy. It will create a dip so steep that once any ball bearings you roll in its direction get too close to it, they'll always fall into the dip and cannot get out no matter how fast they are moving. Around a real black hole, the event horizon is the distance from the black hole where the dip is so steep that not even light can move fast enough to escape. And that's why black holes are black.

There's so much more to tell about the story of black holes. How are they formed? Where

© ESA/Hubble/M. Kornmesser

They're a massive X-ray source

The first observational evidence for a black hole emerged in 1971, coming from a binary star system within our own galaxy. Called Cygnus X-1, the system produces some of the universe's brightest X-rays. These don't emanate from the black hole itself, or from its visible companion star, which is enormous at 33 times the mass of our own Sun. Instead matter is constantly being stripped from the giant star and dragged into an accretion disc around the black hole, and it's from this accretion disc that the X-rays are emitted. As was the case with HR 6819, astronomers used observed star motion to estimate the mass of the unseen object in Cygnus X-1. The latest calculations put the dark object at 21 solar masses, concentrated into such a small space that it couldn't be anything other than a black hole.

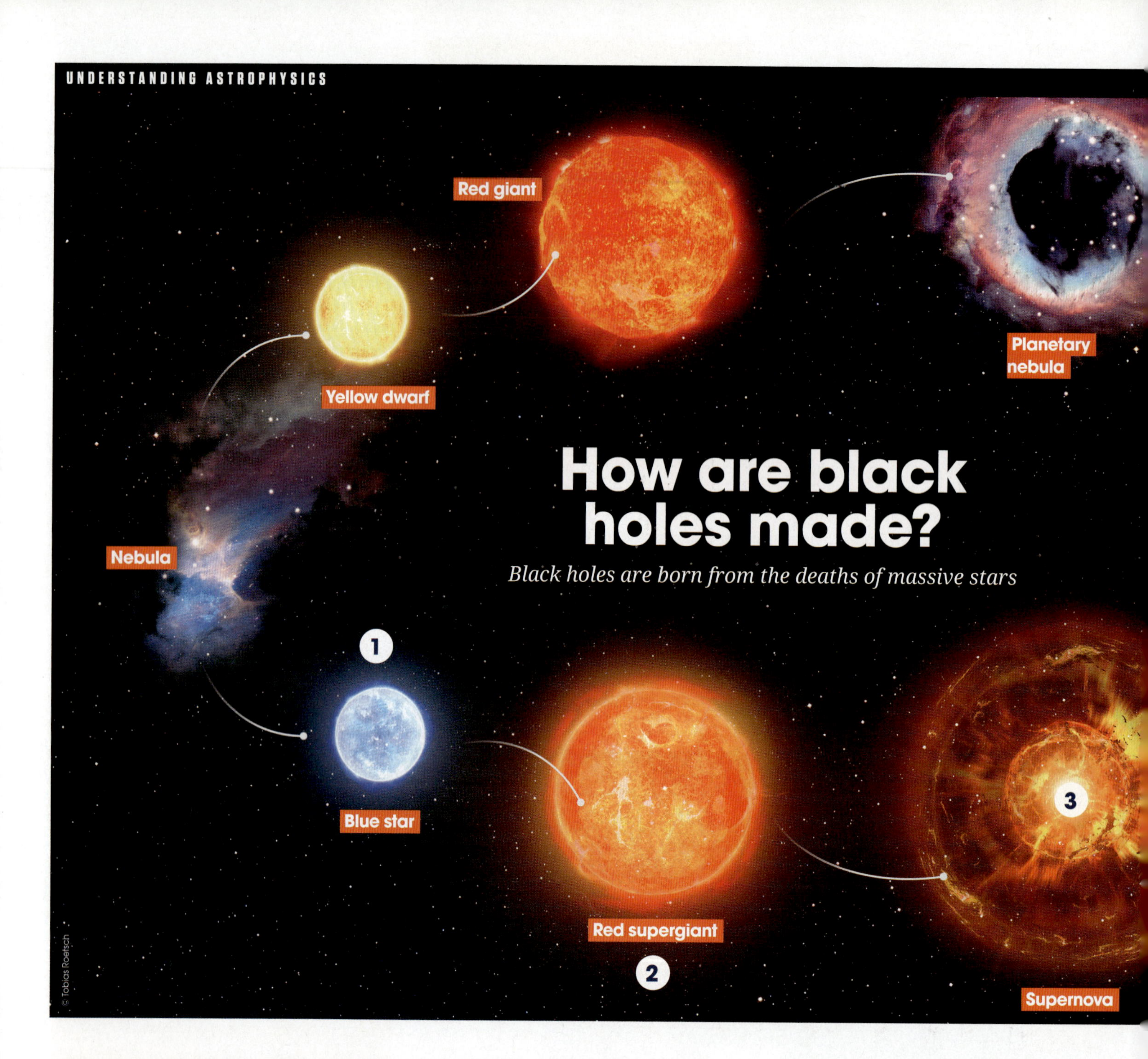

How are black holes made?

Black holes are born from the deaths of massive stars

Advanced space telescopes might be able to image black holes – or at least their surrounding material

do the things that fall into them go? What exists at the centre of a black hole? And what do black holes do in the centres of galaxies? Let's continue with the easier bit – how they are formed. When our Sun, a star, reaches the end of its life in about 5 billion years, it will expand to become a red giant before gracefully puffing off its outer layers to form a planetary nebula, leaving behind its small, hot core, called a white dwarf. Stars more than eight times the mass of the Sun, however, bow out more explosively. Their huge masses ultimately cause their cores to collapse due to the internal pull of their own gravity, while the rest of the star goes supernova. As the outer parts of the star explode, the collapsing core condenses to become a neutron star, which is so tightly packed that it contains as much, or more, mass as the Sun but is only 20 kilometres (12.4 miles) or so across. If the star is massive enough – at least 30 to 40 times the mass of the Sun – then the collapse will continue past the neutron star stage, imploding and collapsing down to a point of infinite density, the 'singularity' at the heart of a black hole.

In mathematics, singularities are calculations that tend to infinity, usually because of some error, some gap in our knowledge needed to fully complete the equation. This also describes the black hole singularity, in the sense that we don't know the necessary physics to figure the singularity out. That's because at the microscopic scale of the black hole singularity, we enter the world of quantum physics, and

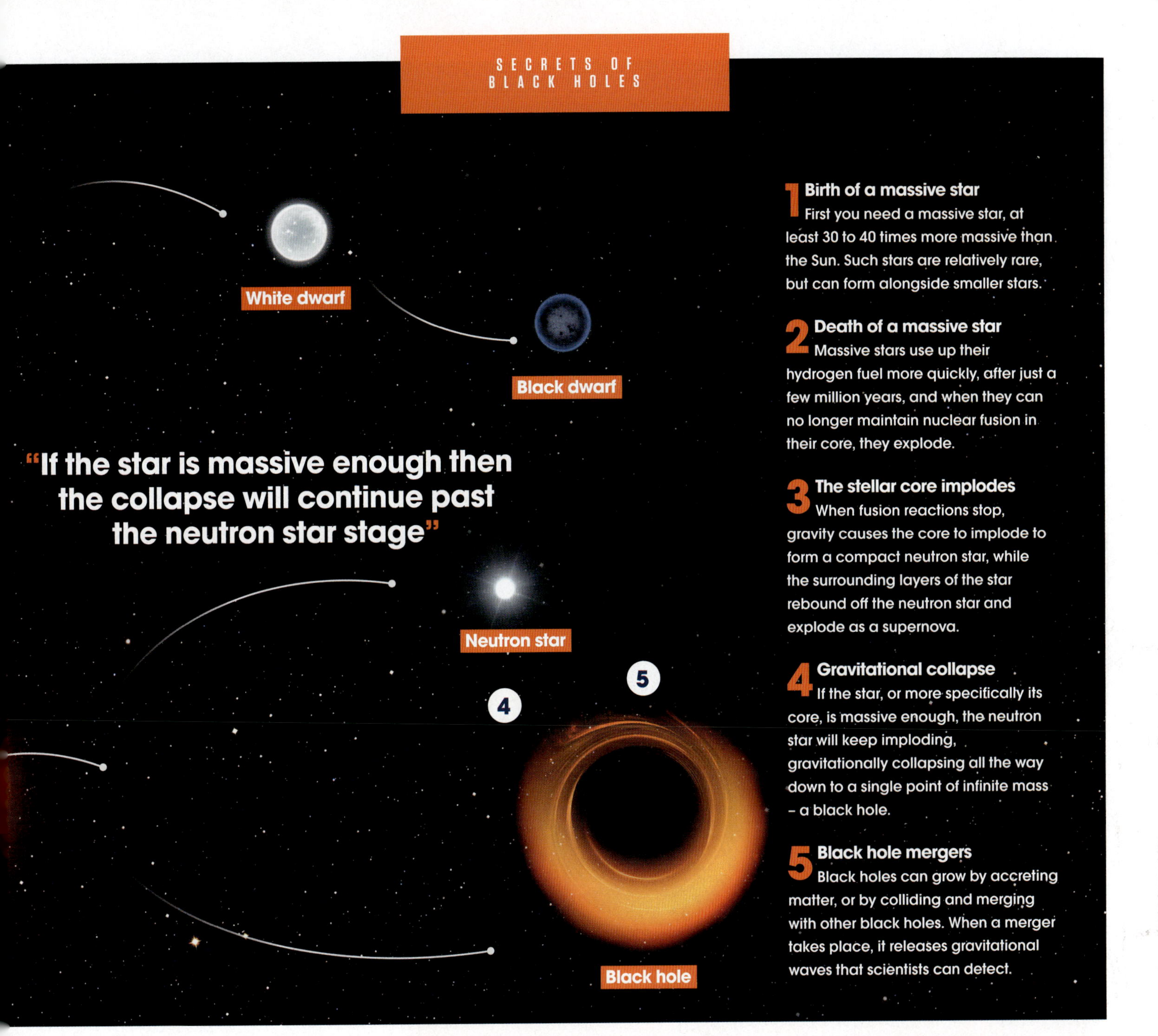

"If the star is massive enough then the collapse will continue past the neutron star stage"

1 Birth of a massive star
First you need a massive star, at least 30 to 40 times more massive than the Sun. Such stars are relatively rare, but can form alongside smaller stars.

2 Death of a massive star
Massive stars use up their hydrogen fuel more quickly, after just a few million years, and when they can no longer maintain nuclear fusion in their core, they explode.

3 The stellar core implodes
When fusion reactions stop, gravity causes the core to implode to form a compact neutron star, while the surrounding layers of the star rebound off the neutron star and explode as a supernova.

4 Gravitational collapse
If the star, or more specifically its core, is massive enough, the neutron star will keep imploding, gravitationally collapsing all the way down to a single point of infinite mass – a black hole.

5 Black hole mergers
Black holes can grow by accreting matter, or by colliding and merging with other black holes. When a merger takes place, it releases gravitational waves that scientists can detect.

scientists don't yet have a theory of quantum gravity. Until we do, scientists won't have the tools to be able to mathematically describe a black hole singularity.

Not that any of this prevents astronomers from learning more about what occurs outside the event horizon, where light still escapes. Some black holes can grow to be millions, or even billions of times more massive than our Sun – in case you're wondering, our Sun has a mass of 1.9 x 1030 kilograms, or 1.9 million trillion trillion kilograms. Astronomers aren't entirely sure how black holes grow to be 'supermassive' like this – it's a possibility that supermassive black holes are the result of lots of mergers of smaller black holes created by supernovae, or maybe they are formed

How black holes grow

The earliest black holes to form will be much bigger today

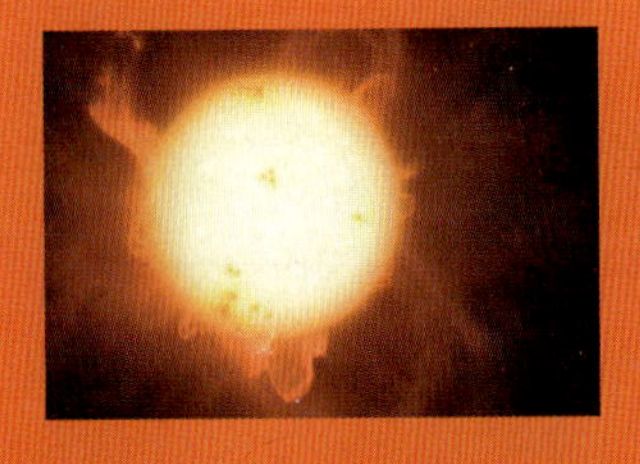

GIANT STAR
The first generation of very massive stars, hundreds of times as massive as the Sun, would have burnt through their nuclear fuel very quickly.

SEED BLACK HOLE
These stars collapsed down to black holes of tens of solar masses, which then acted as seeds for the creation of much larger black holes.

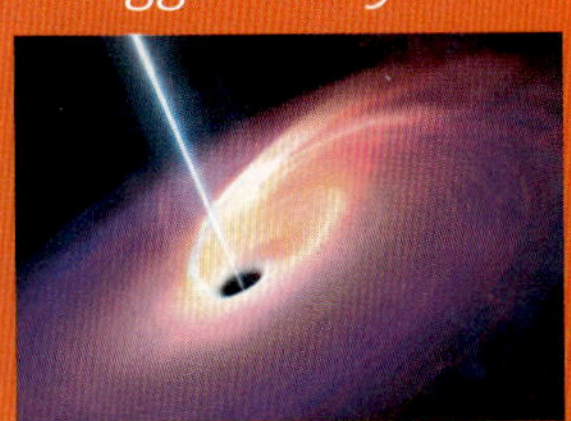

ACCRETION
Over billions of years, gas and dust spiralling into these black holes increased their mass; this isn't enough to explain supermassive black holes.

Anatomy of a black hole

1 Singularity
The core of a black hole is a mysterious point of infinite mass called a singularity, where our currently known laws of physics break down.

2 Event horizon
The point of no return; this is where the gravity from the extreme curvature of space-time is so strong that not even light can escape.

3 Ergosphere
An oblate zone around a rotating black hole, the ergosphere is a volume of space from where energy and mass can be extracted from the black hole.

4 Static limit
The edge of the ergosphere. Everything inside the static limit is caught up by the mass of the black hole dragging the fabric of space-time with it as it rotates.

1
3
2
5
6

They can emit gravitational waves

Black holes don't always exist in isolation; sometimes they occur in pairs, orbiting around each other. When they do, the gravitational interaction between them creates ripples in space-time, which propagate outwards as gravitational waves. With observatories like the Laser Interferometer Gravitational-Wave Observatory (LIGO) and Virgo, we now have the ability to detect these waves. The first discovery, involving the merger of two black holes, was announced back in 2016, and many more have been made since then. As detector sensitivity improves, other wave-generating events besides black hole mergers are being discovered, such as a crash between a black hole and a neutron star which took place way beyond our galaxy at a distance of 650 million to 1.5 billion light years from Earth.

5 Accretion disc
An active black hole siphons and steals matter, mostly in the form of interstellar gas, from the environment around it. This matter falls towards the black hole in a spiralling accretion disc.

6 Relativistic jet
Powerful magnetic fields weaving through the accretion disc can funnel charged particles away in beams or jets that emanate from above and below the rotational axis.

4

"Fortunately, no astronaut has ever fallen into a black hole, but plenty of gas clouds have, and they get spaghettified too"

by giant clouds of gas that existed when the universe was very young which underwent a dramatic gravitational collapse and imploded to form a massive black hole. However they form, supermassive black holes can be found at the centres of most giant galaxies. For example, our Milky Way has a supermassive black hole at its heart called Sagittarius A*, and it has a mass 4.1 million times greater than the Sun's mass.

The huge gravitational well of supermassive black holes means that they can pull in a lot of surrounding material – gas and dust clouds, asteroids, comets and sometimes even whole stars. This material gets ripped apart by the gravitational tidal forces being wielded by the black hole and can lead to a phenomenon known as 'spaghettification'.

Imagine an unfortunate astronaut floating too close to a black hole. The tidal forces are so great that the gravity pulling on the astronaut's feet will be far stronger than the gravity pulling on their head. This would have the effect of stretching them out to the point that they would be pulled into strings of their respective atoms and molecules. Fortunately, no astronaut has ever fallen into a black hole, but plenty of gas clouds have, and they get spaghettified too.

We can see the consequences of these gas clouds being torn apart around active black holes – the material ripped from the gas clouds forms a disc of incredibly hot gas that encircles the black hole outside the event horizon. Astronomers call such discs 'accretion discs' because the gas is said to be accreting onto the black hole. Friction

When black holes collide

Merging galaxies
Two galaxies crash into each other, initially producing a confused mess of material but then settling down to become a single, merged galaxy.

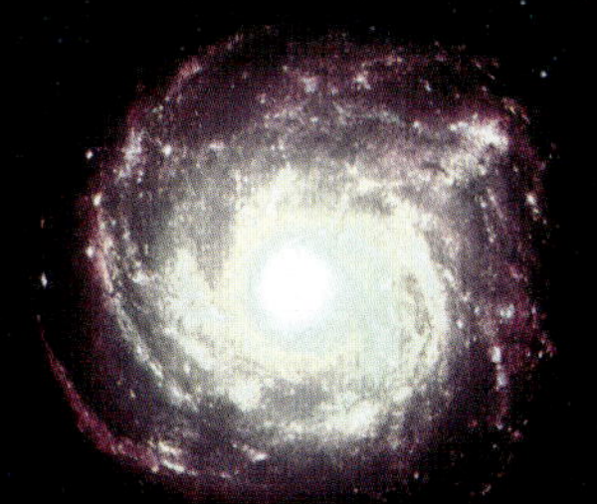

Two black holes
The black holes that started at the centres of the original galaxies gradually spiral towards each other, eventually coalescing.

High-energy jet
The combined black hole may cause the core of the merged galaxy to eject streams of hot gas and high-speed particles.

Visible consequences
The newly merged black hole gives its presence away through the release of gravitational waves and high-energy radiation.

© NASA/CXC/M.Weiss

Gamma-ray bursts are evidence

In the 1930s, Subrahmanyan Chandrasekhar looked at what happens to a star when it has used up all its nuclear fuel. The end result depends on the star's mass. If the star is really big then its dense core – which may itself be three or more times the mass of the Sun – collapses all the way down to a black hole. The final core collapse happens quickly, in a matter of seconds, and it releases a tremendous amount of energy in the form of a gamma-ray burst. This burst can radiate as much energy into space as an ordinary star emits in its entire lifetime. Telescopes on Earth have detected many of these bursts, some of which come from galaxies billions of light years away, so we can actually see black holes being born.

between the atoms and molecules in the disc, which can be moving around the black hole at high speeds, causes the gas to heat up above a million degrees Celsius (1.8 million degrees Fahrenheit) and shine brightly. So while a black hole itself is dark, the environment just outside an active black hole's event horizon can be highly luminous. Furthermore, the disc is rife with powerful magnetic fields emanating from the black hole, and these can funnel charged particles in the disc towards the black hole's rotational axis. The energies are so great that these particles are then magnetically beamed away from the black hole in visible jets that move at almost the speed of light. These jets, and the accretion discs they originate from, are incredibly bright, and when our line of sight is looking almost straight down one of these jets, we see a luminous object called a quasar. When we happen to be looking directly down the jet, it's even more luminous, and we call that a blazar. Regardless of what astronomers name it, the phenomenon is the same – a monster black hole that's spitting out a meal.

Some of the gas in the accretion disc does eventually find itself spiralling into the black hole. For anything approaching and crossing the event horizon, odd things happen. The warping of space-time by the mass of the black hole is so great here that something called gravitational time dilation occurs. As a person approaches the event horizon, time begins to run differently compared to the clocks belonging to observers who are watching from a distance – for example, us watching a black hole with our telescopes. A distant observer would see time stand still at the event horizon, and any astronaut crossing the event horizon would appear nearly frozen in time. The astronaut, however, would not perceive time slowing down; assuming they've somehow survived spaghettification, time will seem to proceed normally to them. It's just one of the weird consequences of general relativity.

And what of material that does fall irrevocably into a black hole? Where does it go, and can it ever come back out again? Stephen Hawking asked this question, and even made a famous bet about it. It turns out that black holes aren't truly black. Hawking became famous for the concept of Hawking radiation, which was the realisation that black holes can actually radiate particles, and even light. The secret lies in quantum field theory, which is a way of saying that on the quantum level, space is continuously fizzing with energy, spontaneously producing pairs of 'virtual' particles – one made of matter, the other of antimatter, such as a positron and an electron. They're described as 'virtual' because they usually instantly annihilate one another, as matter and

Different kinds of black hole

A small black hole

Despite packing in more than 4 million times the mass of our Sun, the black hole at the centre of our galaxy is no larger than the Solar System.

The closest black hole

The nearest known black hole candidate is found in the triple star system HR 6819, which is 1,120 light years from Earth.

Most massive black hole

The heaviest of heavyweights is the supermassive black hole found in the quasar TON 618, with a mass 66 billion times greater than our Sun.

Runaway black holes

Galaxies often collide. When they do, their supermassive black holes also merge, which can give the product of the merger a kick, causing it to escape its galaxy.

Miniature black holes

Some theories suggest that the Big Bang created a horde of micro black holes, with masses ranging from 100 millionths of a kilogram up to that of a small asteroid.

Famous black holes and candidates

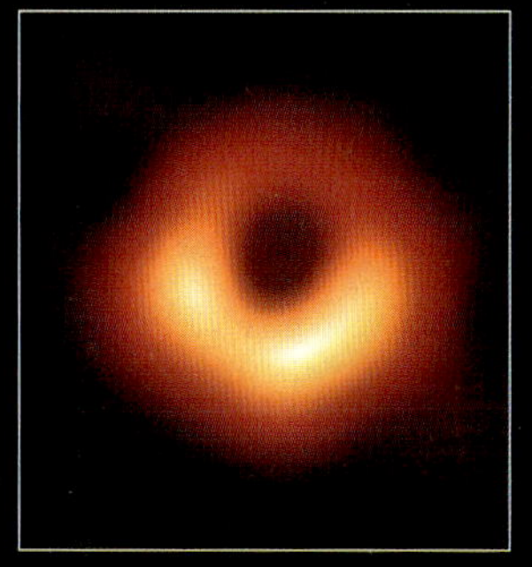

Cygnus X-1

Distance: 6,000 light years

Solar masses: 21.2

A stellar-mass black hole orbiting a blue supergiant star, from which the black hole is stealing gas that forms an accretion disc. It occasionally outbursts in X-rays.

Sagittarius A*

Distance: 26,000 light years

Solar masses: 4.1 million

The supermassive black hole at the centre of our Milky Way. It's generally inactive, with only modest X-ray outbursts as it consumes small gas clouds.

Messier 87 black hole

Distance: 54 million light years

Solar masses: 6.5 billion

The first black hole to be imaged right down to the event horizon, revealing the black hole's 'shadow' on the surrounding accretion disc.

GW150914 black hole

Distance: 1.4 billion light years

Solar masses: 62

The product of the first black hole merger to be detected by gravitational waves formed when a 35-solar-mass black hole collided with a 30-solar-mass black hole. The extra three solar masses were converted into gravitational-wave energy.

3C 273

Distance: 2.4 billion light years

Solar masses: 886 million

The first quasar discovered, the black hole at its heart is hungrily guzzling gas, producing an incredibly bright accretion disc and a jet moving at almost the speed of light.

> "Some of the gas in the accretion disc does eventually find itself spiralling into the black hole"

The Atacama Large Millimeter/submillimeter Array (ALMA) searches for signals of black holes

The network that imaged a black hole

These global telescopes blazed a trail for black hole imaging and revolutionised astrophysics

1 South Pole Telescope
Location: Antarctica

2 Arizona Radio Observatory's Submillimeter Telescope
Location: Arizona

3 James Clerk Maxwell Telescope
Location: Hawaii

4 Submillimeter Array
Location: Hawaii

5 Large Millimeter Telescope
Location: Mexico

6 Atacama Large Millimeter /submillimeter Array
Location: Chile

7 Atacama Pathfinder Experiment
Location: Chile

8 Very Long Baseline Array
Location: New Mexico

9 Robert C. Byrd Green Bank Telescope
Location: West Virginia

10 Radio Telescope Effelsberg
Location: Germany

11 Yebes Observatory
Location: Spain

12 Northern Extended Millimeter Array
Location: France

13 Institut de Radioastronomie Millimétrique
Location: Spain

14 Combined Array for Research in Millimeter-wave Astronomy
Location: California

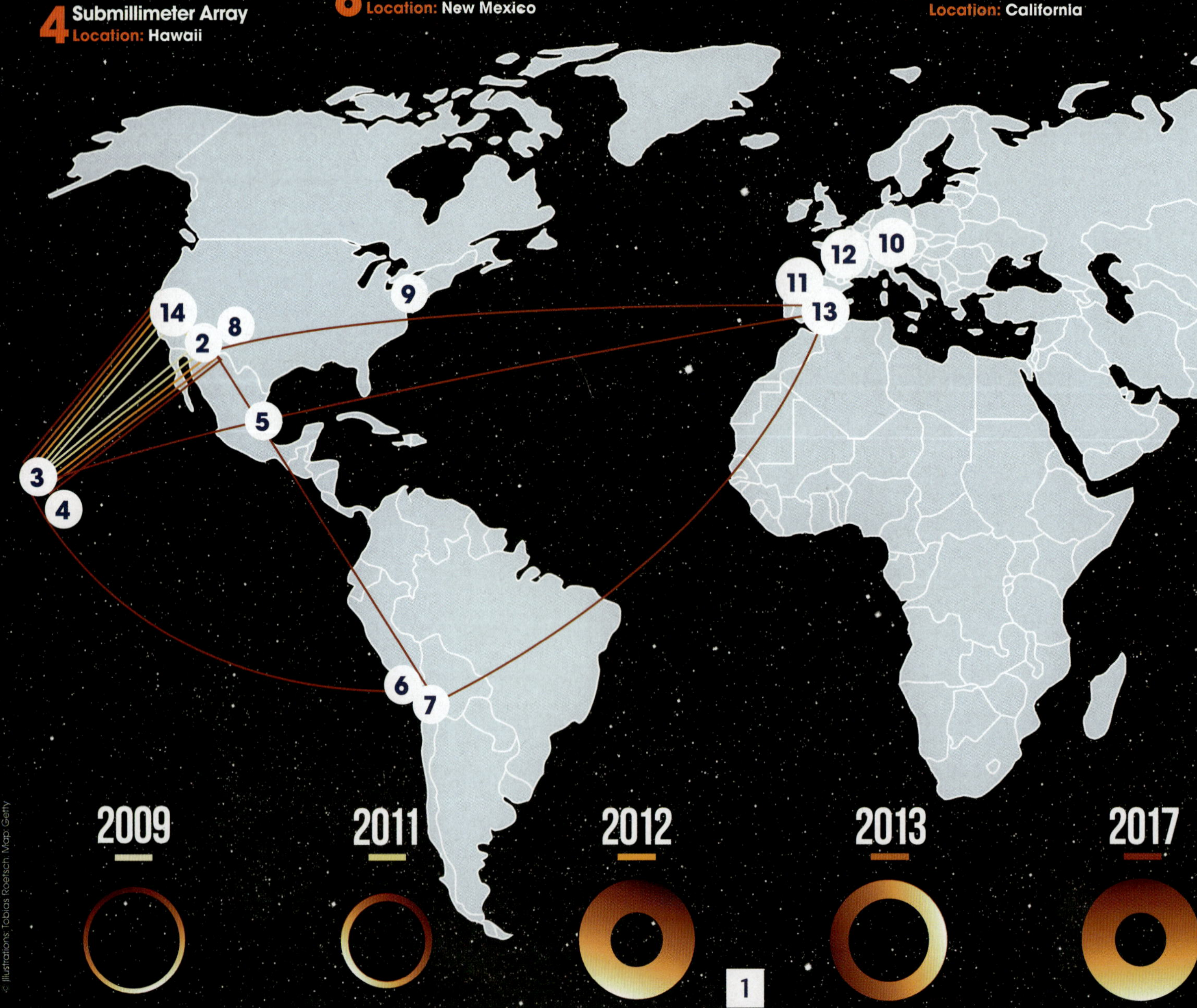

© Illustrations: Tobias Roetsch, Map: Getty

How the Event Horizon Telescope works

The planet-sized array used cutting-edge technology to reveal the edge of a black hole

Cloaking Device

Everything inside the event horizon of a black hole is forever hidden from view, because not even light can escape. But just outside this, the matter spiralling inwards shines brightly.

Gravitational lens

Light that passes close to the event horizon gets deflected, as if it was passing through a lens. This warps the circular accretion disc from our point of view.

Glowing banana

Matter orbits so fast close to the black hole that the radiation shifts wavelength from one side to the other, making it brighter or darker to us.

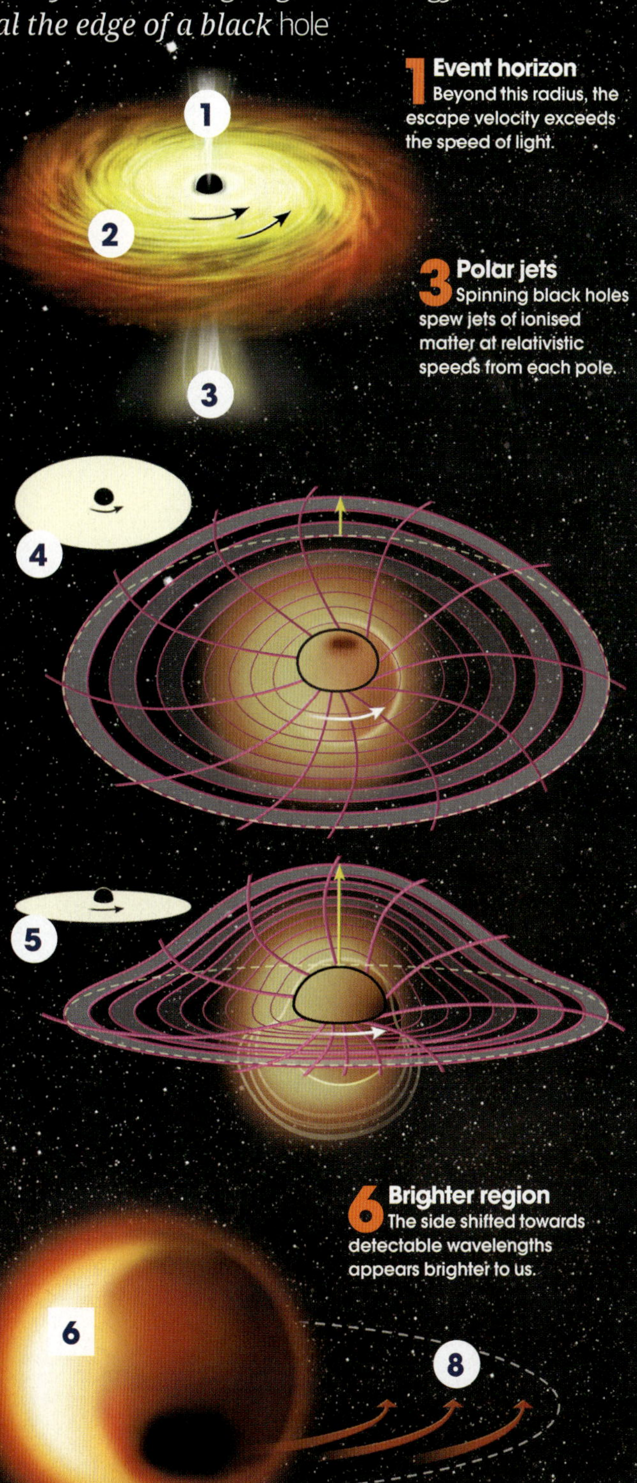

1 Event horizon Beyond this radius, the escape velocity exceeds the speed of light.

2 Accretion disc Dust and gas gets accelerated almost to the speed of light.

3 Polar jets Spinning black holes spew jets of ionised matter at relativistic speeds from each pole.

4 Face on If the accretion disc is nearly perpendicular to us, the gravitational lens effect is small.

5 Edge on At oblique angles, the image is bent upwards, showing us behind the black hole.

6 Brighter region The side shifted towards detectable wavelengths appears brighter to us.

7 Blueshift Light moving towards Earth gets squashed towards shorter wavelengths.

8 Redshift Light moving away from Earth is stretched into longer wavelengths.

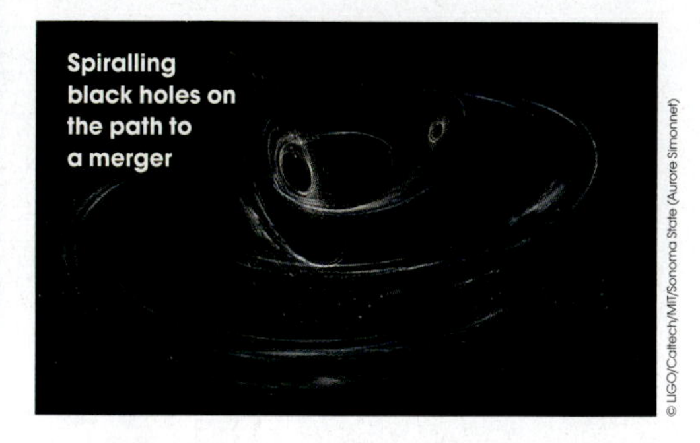
Spiralling black holes on the path to a merger

antimatter do when they come into contact with one another, so they're not in existence for very long. But Hawking realised that pairs of virtual particles coming into existence on the edge of the event horizon can be split up – one falls into the black hole, while the other, if it has enough energy, can race away into space and escape. Since the escaping particle has lost its antipartner, it doesn't annihilate, surviving to become a 'real' particle. Since the energy of the quantum field is drawn from the black hole's mass, the escaping particle is essentially running off with some of the mass of the black hole. Over the course of trillions and trillions of years, even the most supermassive black holes will begin to evaporate through the release of Hawking radiation.

Once a black hole evaporates, what happens to all the information of everything that went into it? This is known as the black hole information paradox, which became the focus of Hawking's famous bet with fellow physicists Kip Thorne and John Preskill. They would often make bets with one another, and in this particular case Hawking and Thorne bet Preskill that information inside a black hole is not preserved. In 2004 Hawking conceded by agreeing that black holes do preserve information through Hawking radiation and gave Preskill a baseball encyclopaedia, 'made for the storage and retrieval of information'.

Black holes fascinate us because they are so far out of our everyday experience, and there's still so much we don't know about them. When the Event Horizon Telescope took the very first image of a black hole's event horizon, it made the front pages of newspapers. Who knows what new and bigger telescopes will discover about black holes in the future?

Giles Sparrow
Space science writer
Giles has degrees in astronomy and science communication and has written many books and articles on all aspects of the universe.

Everything you need to know about the Solar System

Join us on a tour through our current understanding of the planetary system we call home

Our Solar System consists of the area influenced by the Sun and, apart from occasional stray visitors from interstellar space, everything it contains. Aside from the Sun, its main components are the eight major planets, their moons and rings, a handful of worlds classified as dwarf planets and vast numbers of smaller bodies made of varying amounts of rock and ice, which are broadly termed asteroids and comets. Most of these objects orbit in a plane roughly in line with the Sun's equator and in the same direction as the Sun's rotation – anticlockwise when viewed from 'above' the plane.

The four innermost planets are mostly composed of dense rock and metal. Earth is the largest of these 'terrestrial' planets, with Venus almost the same size, Mars significantly smaller and Mercury the smallest of all. A large gap separates the orbit of Mars from that of Jupiter, the innermost gas giant and the largest planet in the entire Solar System, with a diameter of 11.2 Earths. Saturn is somewhat smaller, and outer Uranus and Neptune are near twins, both about four times the diameter of Earth.

The entire Solar System sits in the Milky Way – a vast spiral galaxy within which the Sun is just one of several hundred billion stars. At about 26,000 light years from the centre, it takes some 230 million years to complete one trip around the galaxy.

An evolving system

Although today's Solar System seems stable, it represents just a snapshot in a long history of change and evolution. Asteroids and comets in planet-crossing orbits are doomed to suffer disruption of some kind on astronomical timescales, and so their supplies must continuously be replenished. In the first billion years of Solar System history, however, changes were far more dramatic. It's increasingly clear that the giant planets formed closer to the Sun – and to each other – than they are now, and a subsequent gravitational tug of war saw their orbits evolve and change. Jupiter may have first migrated even closer to the Sun, scattering vast numbers of icy objects from the outer edge of today's asteroid belt and beyond into extreme elliptical orbits to form the Oort Cloud, before reversing its track. Neptune may have started its life closer to the Sun than Uranus before their own complex gravitational dance swapped them over, pulling Uranus' axis of rotation over to its extreme 98-degree angle. Some computer models even suggest that in order to reach the current configuration of giant planets, there must once have been a fifth Neptune-sized world that was long ago ejected from the Solar System entirely, or perhaps flung into exile amid the comets of the Oort Cloud.

Origins

Evidence from rock grains in ancient asteroids suggests the Solar System began to form about 4.57 billion years ago. Like other stars, the Sun was born from a collapsing cloud, or nebula, rich in gas and dust. As the centre of the cloud grew hotter and denser it began to spin more rapidly, while material around it flattened out into a rotating disc. Dust grains collided and stuck together in the disc, perhaps growing step by step through chance collisions until they had sufficient gravity to draw in more material from around them, or perhaps forming huge clouds of orbiting 'pebbles' that underwent sudden collapse into larger protoplanets when they became unstable.

Meanwhile, as the Sun became hot enough to shine properly, rising temperatures caused easily melted chemicals to evaporate as far out as an 'ice line' in the present-day asteroid belt. Simultaneously, fiery radiation from the newborn Sun and a solar wind of ionised particles blowing out from its surface began to drive gas outwards. While the worlds of the inner Solar System had to form mostly from dry, rocky materials, those farther out incorporated substantial amounts of ice, and in the case of the largest planets were also able to keep hold of huge gaseous atmospheres thanks to their powerful gravity.

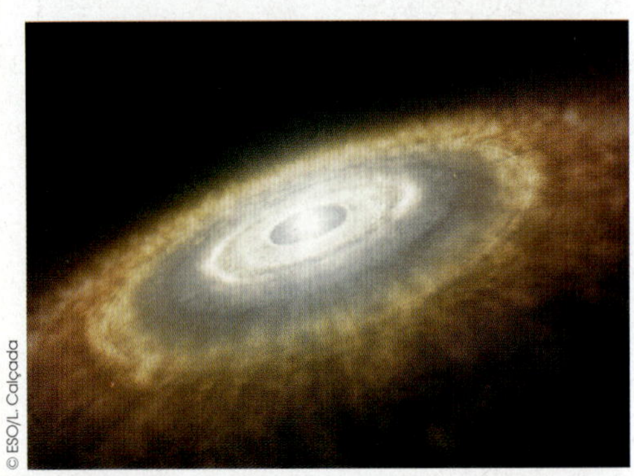

Everything in the Solar System formed from a protoplanetary disc around the Sun

The Sun sends out ionised particles that carry through to the outer reaches of the Solar System

The Sun

Our local star controls conditions across the wider Solar System. With a visible diameter of 1.39 million kilometres (860,000 miles), it accounts for some 99.8 per cent of the Solar System's entire mass and has a composition dominated by hydrogen – the lightest and simplest gas in the universe.

The Sun shines by nuclear fusion, a process that forces hydrogen nuclei together in the core to form nuclei of helium, the next lightest element. Energy is released in the process as photons of high-energy radiation that gradually force their way outwards through the overlying layers, losing energy as they do, and keeping the Sun's interior hot. The Sun's incandescent visible surface, or photosphere, marks the region where its gas becomes cool and sparse enough to be transparent, and visible, infrared and ultraviolet light can escape. This surface has an average temperature of around 5,500 degrees Celsius (9,932 degrees Fahrenheit), although dark sunspots, created where the Sun's tangled magnetic field bursts from its surface, can be a couple of thousand degrees cooler.

Above the photosphere, the Sun's upper layers are home to violent activity that varies,

"As the Sun became hot enough to shine properly, rising temperatures caused easily melted chemicals to evaporate"

Our home planet is the largest of the four rocky bodies closest to the Sun

along with sunspot numbers, in an 11-year cycle. The cycle significantly affects the shape of the Sun's corona, or outer atmosphere, which typically extends to several times its visible diameter before merging with the solar wind of particles blowing out across the Solar System.

Rocky planets

Many factors have shaped the evolution of the terrestrial planets – most importantly their size, composition and distance from the Sun. As a rule, the larger a planet is, the hotter its interior will remain, giving rise to a more complex structure and potentially a molten metallic core. Size and mass determine a planet's gravity, which along with its temperature and the presence of a protective magnetic field influence how well it can hold on to an atmosphere. These factors influence the chemicals that can exist on its surface.

All four rocky planets were likely bombarded by icy objects from farther out in the Solar System during or shortly after their formation, returning water to their surfaces. Venus, Earth and Mars all once had oceans of liquid water, but Venus' was lost to a runaway greenhouse effect early in its history, leaving behind an arid, hellish landscape. The weak gravity and lack of a protective magnetic field around Mars allowed much of its atmosphere and water to escape into space, cooling the surface until most of the remaining water became locked in permafrost and the polar ice caps. Venus and Mars show signs of geological activity in the relatively recent past, but this mostly takes the form of volcanism, while activity on Earth is far more complex and continuous.

Giants of gas and ice

The giant planets of the outer Solar System are broadly divided into two pairs: the inner gas giants Jupiter and Saturn, dominated by huge envelopes of hydrogen, and the outer ice giants Uranus and Neptune, made of more complex chemicals such as water, methane and ammonia. All four have deep outer atmospheres that are home to complex weather systems. Despite their size these planets spin rapidly, generating high winds that wrap cloud systems into bands parallel to their equator.

Beneath the active atmospheres of Jupiter and Saturn, pressure from above forces hydrogen into a liquid state, and can even break it down into liquid metallic form, generating extremely powerful magnetic fields. The deeper layers of Uranus and Neptune, meanwhile, are composed of icy chemicals in liquid form. Slow contractions of the inner layers due to gravity, coupled with chemical reactions, generates significant heat inside three of the giants, though Uranus is a mysterious exception, helping to power their weather systems even in the cold outer Solar System.

The considerable gravity of the giants puts each one at the centre of its own substantial satellite system – all four are orbited by a mix of 'regular' moons, formed from material left in orbit as the planet itself formed, and 'irregular' objects captured during later close encounters. Each giant also has a ring system of its own, made up of particles trapped in concentric orbits. These vary wildly between the broad, icy planes of Saturn to the tenuous dust around Jupiter and the tightly defined arcs around Uranus and Neptune.

Ceres

Dwarf planets

The term dwarf planet was introduced to clarify the organisation of the Solar System in 2006 – though some might say it's made matters more confusing. Dwarf planets are worlds in orbit around the Sun with sufficient gravity to pull themselves into a spherical shape, but not enough to deflect the paths of other nearby bodies and 'clear their orbits'.

The first dwarfs to be discovered were Ceres in 1801 and Pluto in 1930. Both were originally treated as new major planets, despite their small size, but Ceres was swiftly reclassified as

Pluto

Beyond the asteroid belt, the planets are gaseous and gigantic in comparison to Earth

Life in the Solar System

Earth's abundant life is due to its unique position in the Solar System near the inner edge of our Solar System's 'Goldilocks zone', where temperatures are neither too hot nor too cold, but 'just right' for liquid water to survive on a planet's surface. Water is widely seen as a key requirement for life because it's the most abundant and effective 'solvent' that we know of – a chemical within which other molecules can dissolve and move around, permitting the encounters and reactions that are needed for life to evolve and survive.

Mars is the only other planet technically just within the Goldilocks zone, and its warmer, wetter history makes it an intriguing destination in the search for past or present life, but there are also surprising possibilities farther from the Sun.

Several large satellites and dwarf planets seem to have liquid-water layers deep in their interiors, while tidal forces raised by Jupiter and Saturn on their icy moons Europa and Enceladus pummel and heat their interiors so much that they have substantial liquid-water oceans just below the surface. Fed with chemical nutrients by undersea volcanoes, these two worlds are seen as the Solar System's most likely spots for alien life to exist.

© Alamy

There are millions of tiny icy and rocky fragments floating through space

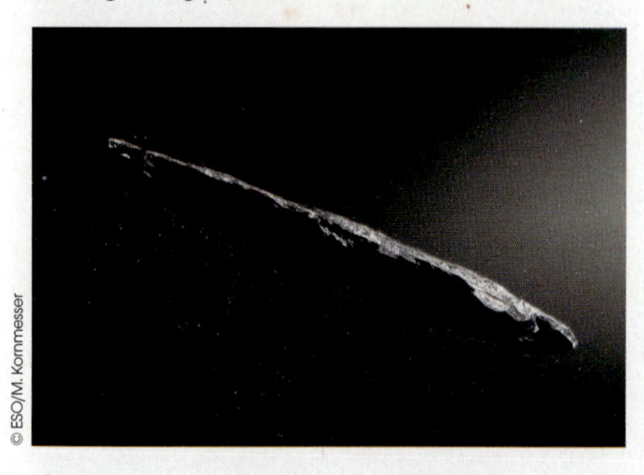
© ESO/M. Kommesser

'Oumuamua was identified as an interstellar interloper passing through our Solar System

an asteroid once more of its neighbours in the main asteroid belt were discovered. Pluto's status became doubtful in the 1990s as more small bodies in similar orbits were found in the Kuiper Belt, but matters came to a head with the discovery of Eris, another 'trans-Neptunian object' of similar size, in 2003.

Faced with a potentially ballooning list of 'major' planets, astronomers opted to introduce the new category, demoting Pluto, but sweeping up Ceres into the bargain. Because dwarf planets are classified in part by their shape, and this is still uncertain for some distant worlds, there are still fierce debates about which objects qualify. The International Astronomical Union currently recognises just five: Ceres, Pluto, Haumea, Makemake and Eris.

Rocky debris

Although the formation of the Solar System left plenty of rock and dust scattered across the inner Solar System, most of the smaller rocky objects that survive today are confined to the asteroid belt between Mars and Jupiter, where the giant planet's gravity and early shifts in its orbit disrupted any potential for the formation of a fifth rocky planet. Today's asteroid belt contains around 1.5 million asteroids more than one kilometre (0.6 miles) across, along with countless smaller objects.

Although they're scattered across such a vast volume of space that crossing the belt is easy, collisions and close encounters are inevitable on a longer timescale. These lead to the formation of asteroid families with similar compositions and orbits that can be traced back to a common origin. Asteroids vary in composition from 'carbonaceous' objects that have barely altered since the birth of the Solar System to bodies rich in silicate minerals or even iron – fragments of larger ancient worlds that had begun to develop an internal structure before they were smashed apart.

Collisions can also send asteroids onto elliptical orbits that cross over those of the inner planets, with some becoming potentially hazardous near-Earth objects, or NEOs. However, NEO orbits are inevitably unstable over long timescales – ending either in a collision with a major planet or more likely deflection from a close encounter – and so this supply must be steadily renewed.

Icy wanderers

The farther out we look in the Solar System, the more volatile ices – not just water ice, but also frozen methane and other compounds – become mixed with the rocky components of solid bodies. This trend is already apparent fairly close to the Sun in the asteroid belt, but it becomes more pronounced among the moons of the giant planets, and above all in the small worlds of the Kuiper Belt beyond Neptune.

The most familiar icy objects, however, are comets. These icy wanderers spend most of their lives in a deep-frozen state, orbiting among the Kuiper Belt objects or even farther out in the Oort Cloud – a vast, spherical comet cloud that surrounds the Solar System. However, they spark into life when chance puts them on an elliptical orbit that brings them close to the Sun. As the comet's solid nucleus warms up, gases evaporating from the surface first form a vast, diffuse atmosphere, called a 'coma', and then an elongated tail that is caught up on the solar wind and dragged away from the Sun.

Comets that visit the inner Solar System may follow orbits that vary from just a few years to tens of thousands. However, each successive visit strips away some of their ice until they eventually become dark, dormant and – depending on their orbits – barely distinguishable from asteroids.

Testing the limits

Many astronomers from across the world define the Solar System's outer limit as the boundary where the Sun ceases to be the exclusive dominant influence over nearby objects. According to this definition, the edge of the Solar System lies at the heliopause – the wall where the solar wind streaming out from the Sun comes to a halt in the face of pressure from countless other stellar winds and the 'interstellar medium' – clouds of sparse gas that lie between the stars.

This boundary lies around four times farther from the Sun than Neptune, or 120 times farther out than Earth. Four spacecraft – Pioneers 10 and 11 and Voyagers 1 and 2 – have crossed it so far, and the two Voyagers continue to send back data about conditions on the other side.

Despite the widespread adoption of the heliopause as the formal 'edge' of the Solar System, there are many objects in the space beyond it that still orbit the Sun. Most of these lie within either the scattered disc, a broad outer extension of the Kuiper Belt, or the Oort Cloud. According to the most generous definition, the Solar System extends to the edge of the Oort Cloud, roughly a light year from the Sun.

The eight major planets

Mercury
Diameter: 4,879 kilometres (3,032 miles)
Mass: 0.055 Earths
Distance from the Sun: 46 to 69.8 million kilometres (28.6 to 43.4 million miles)
Orbital period: 88 days
Rotation period: 58.65 days
Axial tilt: 0.03 degrees
Satellites: Zero

Venus
Diameter: 12,104 kilometres (7,521 miles)
Mass: 0.815 Earths
Distance from the Sun: 107.5 to 108.9 million kilometres (66.8 to 67.7 million miles)
Orbital period: 224.7 days
Rotation period: 243.02 days
Axial tilt: 177.36 degrees
Satellites: Zero

Earth
Diameter: 12,742 kilometres (7,918 miles)
Mass: 5.97 billion trillion tonnes
Distance from the Sun: 147.1 to 152.1 million kilometres (91.4 to 94.5 million miles)
Orbital period: 365.256 days
Rotation period: 23 hours and 56 minutes
Axial tilt: 23.44 degrees
Satellites: One

Mars
Diameter: 6,779 kilometres (4,212 miles)
Mass: 0.107 Earths
Distance from the Sun: 206.7 to 249.2 million kilometres (128.4 to 154.8 million miles)
Orbital period: 686.98 days
Rotation period: 24 hours and 37 minutes
Axial tilt: 25.19 degrees
Satellites: Two

Jupiter
Diameter: 139,822 kilometres (86,881 miles)
Mass: 317.8 Earths
Distance from the Sun: 740.5 to 816.6 million kilometres (460 to 506.4 million miles)
Orbital period: 11.86 years
Rotation period: 9 hours and 55 minutes
Axial tilt: 3.13 degrees
Known satellites: 79

Saturn
Diameter: 116,464 kilometres (72,367 miles)
Mass: 95.2 Earths
Distance from the Sun: 1.35 to 1.51 billion kilometres (838 to 938 million miles)
Orbital period: 29.46 years
Rotation period: 10 hours and 34 minutes
Axial tilt: 26.73 degrees
Known satellites: 82

Uranus
Diameter: 50,724 kilometres (31,518 miles)
Mass: 14.5 Earths
Distance from the Sun: 2.74 to 3.01 billion kilometres (1.7 to 1.87 billion miles)
Orbital period: 84.02 years
Rotation period: 17 hours and 14 minutes
Axial tilt: 97.77 degrees
Known satellites: 27

Neptune
Diameter: 49,244 kilometres (30,598 miles)
Mass: 17.1 Earths
Distance from the Sun: 4.44 to 4.54 billion kilometres (2.76 to 2.82 billion miles)
Orbital period: 164.8 years
Rotation period: 16 hours and 7 minutes
Axial tilt: 28.32 degrees
Known satellites: 14

Sun
Diameter of photosphere: 1.39 million kilometres (863,706 miles)
Mass: Around 330,000 Earths
Rotation period: 25 days at equator, 34.4 days at the poles

Does the Sun stay still in the Solar System?

We tend to think of our star as the stationary centre of our system, but the planets do make it wobble

Exoplanet hunting has been dominated by NASA's Kepler and Transiting Exoplanet Survey Satellite, which stare out at the stars looking for drops in brightness as the members of other planetary systems pass between ourselves and their parent stars. But the first exoplanets were discovered by the 'wobble' method, which looks for changes in starlight as the star is slowly dragged around by its accompanying planets. The difference is small, as stars are generally much larger than their collection of planets, and for multi-planet systems the motion is a combination of the influence of all partners.

All the bodies involved in an orbital relationship will be influenced by each other to some degree; it's easiest to envisage the most extreme case, two equally sized objects orbiting each other. If Earth and the Moon were the same size, they would orbit each other around a point halfway between them – this is called the 'barycentre', the centre of mass of the whole system. As one body gets larger the barycentre moves closer to it, with the smaller body following a larger circle and the heavier body a smaller one. In reality, Earth is four times the size of the Moon, so the barycentre is 4,670 kilometres (2,902 miles) from Earth's centre, or 1,710 kilometres (1,063 miles) underground. However, in the case of Pluto and Charon, they are so closely matched that their barycentre is 960 kilometres (597 miles) above Pluto's surface.

Expert: Robin Hague

Robin is a science writer, focusing on space and physics. He is head of launch at Skyrora, coordinating launch opportunities for Skyrora's vehicles.

"All the bodies involved in an orbital relationship will be influenced by each other to some degree"

The Sun is also moving through the galaxy, orbiting the Milky Way's centre

So two bodies follow simple circular paths around the barycentre, but the Solar System has at least nine major bodies, depending upon definitions, and all those influences add up. By far the biggest solar wobble is created by Jupiter; it's 2.5 times the mass of all the other planets combined and endeavours to pull the Sun around in one large wobble, centred 46,000 kilometres (28,583 miles) above the nominal solar surface every 11 years. But then Saturn has a mass 30 per cent that of Jupiter and an orbit of 29.5 years, so its wobble is superimposed on Jupiter's, leading to a loop-in-a-circle-shaped path that progressively changes as the planets move positions. Uranus and Neptune are also big enough to introduce wobbles that are significant on the scale of the Sun's radius, and then all the terrestrial and dwarf planets and their moons add their own small wobbles on top.

Ultimately, the Sun dances around a combined path of loops as the planets shift, such that the barycentre of the whole Solar System is sometimes within the solar sphere and sometimes outside. If there should be an alien civilisation out in one of the many systems we now know about, they would be able to tell from analysing the motion of our Sun – as perhaps we have already done with theirs – that there are eight or more planets in its possession.

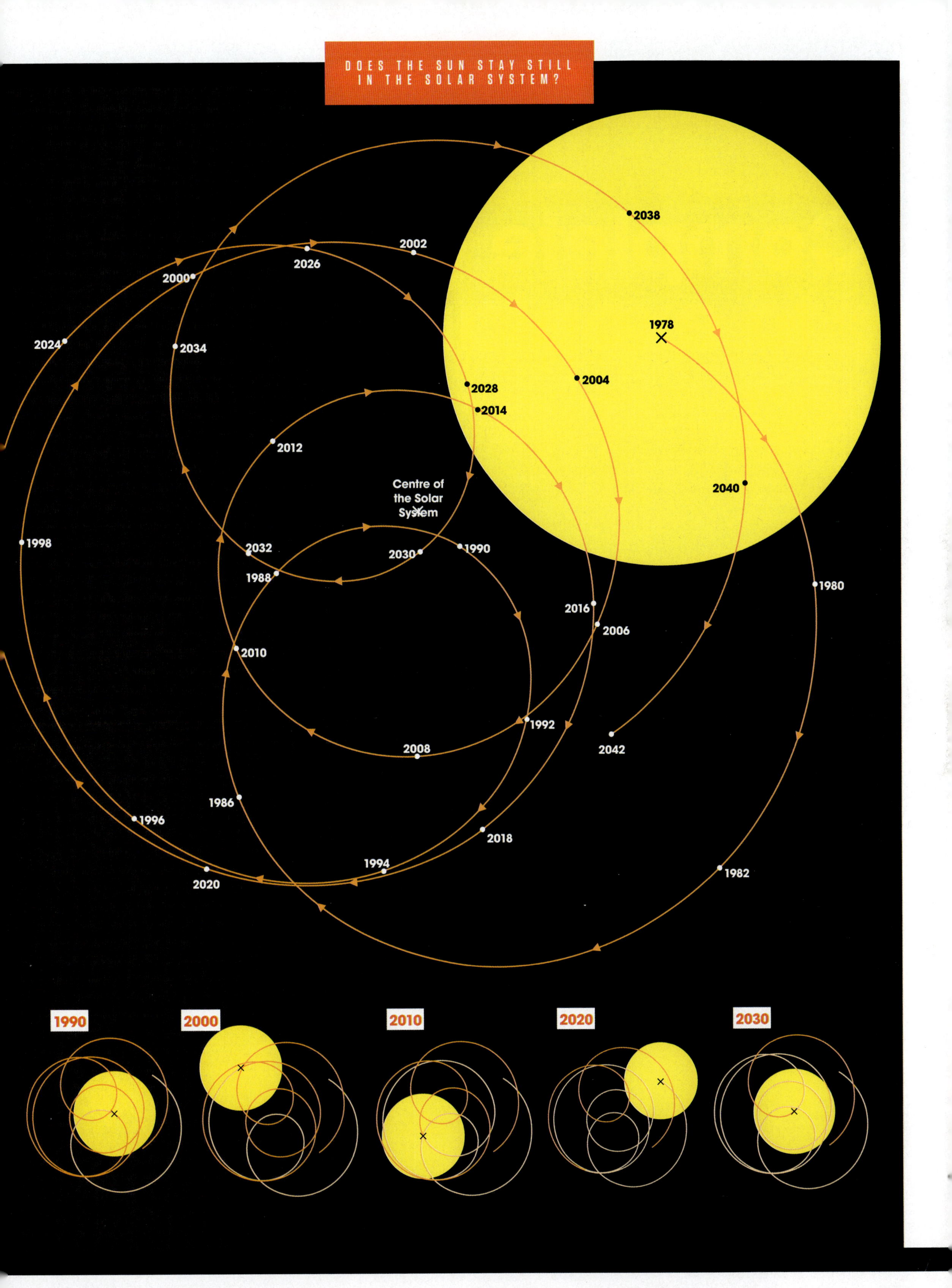
2038
2002
2026
2000
1978
2024
2034
2004
2028
2014
2012
Centre of the Solar System
2040
1998
2032
2030
1990
1988
1980
2016
2006
2010
1992
2008
2042
1986
1996
2018
1994
2020
1982
1990
2000
2010
2020
2030

Complete guide to exoplanets

Our knowledge of worlds beyond the Solar System has exploded in the last three decades

© Tobias Roetsch

Ever since humans first discovered that the stars in the night sky were bodies similar to our Sun, we've dreamed and speculated about the worlds that could orbit these stars. Would they be rocky terrestrial planets like Earth? Could they possess liquid water? Could the presence of this vital life-sustaining element on other worlds mean that we are not alone in the universe?

"For millennia, humans have been asking the question of whether we are alone. And tied to that question, are other planets anywhere else?" Nikku Madhusudhan, a professor of astrophysics and exoplanetary science at the Institute of Astronomy of the University of Cambridge, tells All About Space. "It's very fundamental to being human to ask the question if there are planets elsewhere."

With this considered, it's almost shocking to think that before the 1990s astronomers weren't even certain that stars outside the Solar System possessed their own planets. There was no evidence to suggest that extrasolar planets, or exoplanets for short, didn't exist, nor were there hints that the Solar System was in any way unique in the Milky Way. But until the very end of the 20th century, astronomers had been frustrated by the lack of direct evidence of worlds beyond the influence of our star.

This is because exoplanets are notoriously difficult to detect. Historically, the most successful exoplanet detection methods have worked by inferring the tiny effects that planets have on their parent stars, like tiny dips in light or the near-imperceptible 'wobble' they cause in their star's motion. "Until 30 years ago, we didn't know of any planets outside the Solar System; all we knew of were the planets in the Solar System," Madhusudhan continues.

Types of exoplanet

Hot Jupiters

Mass: Up to 12 times Jupiter's mass

Size: 0.3 to 10.0 times Jupiter's radius

Number discovered: 1,458

Gas giants like Jupiter. The difference is these worlds orbit closer to their stars, with short orbits and blistering surface temperatures. WASP-76b is a planet so close to its host that it completes an orbit in under two days.

Super-Earths

Mass: Up to ten times Earth's mass

Size: Between 0.8 and 4.0 times Earth's radius

Number discovered: 604

Rocky terrestrial worlds or gas planets more massive than Earth but smaller than Neptune. One example of a super-Earth is Gliese 15 A b, a rocky world 11 light years away which is over three times the size of Earth.

Sub-Neptunes

Mass: Up to 17 times Earth's mass

Size: Over 2.0 times Earth's Radius

Number discovered: 1,719

Planets similar in size to the Solar System's ice giant Neptune, they're believed to be the most common type of planet in the Milky Way. Discovered in 2018, Kepler-1655 b has a radius around 2.3 times that of Earth, with five times our planet's mass.

Terrestrial

Mass: Around that of Earth

Size: Between 0.5 to 2.0 times Earth's radius

Number discovered: 186

These are small, rocky worlds like the Solar System's inner planets Mercury, Venus, Earth and Mars. TRAPPIST-1e is the fourth planet from its star, 40 light years from Earth. Existing in the habitable zone of its host, it could potentially have liquid water.

Terrestrial planets have rocky surfaces, like Earth

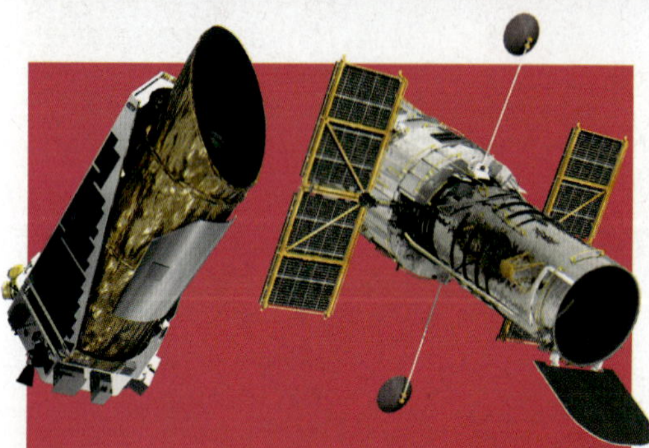

Past exoplanet missions

Hubble Space Telescope

1990-present, 2006 SWEEPS project

The Sagittarius Window Eclipsing Extrasolar Planet Search (SWEEPS) was a survey that used Hubble to detect exoplanets via the transit method. Hubble observations have also helped astronomers study the candidate planets spotted by other telescopes.

Kepler Space Telescope

2009-2018, retired

Kepler discovered over 2,000 confirmed exoplanets, along with thousands more unconfirmed candidates.

"But as soon as exoplanets were discovered, that opened an entirely new window into the universe and its other planetary systems."

Since this point, improved technology and detection techniques have resulted in a bulging exoplanet catalogue of over 4,800 distant worlds. "The first big milestone in the study of exoplanets was the realisation of just how common they are," adds Madhusudhan, who developed a technique of atmospheric retrieval to infer the compositions of exoplanets. "But also, those exoplanets are extremely diverse. Exoplanets come in all sorts of masses, sizes and temperatures."

When it comes to the categorisation of these objects, humanity's Solar System bias is evident. Worlds outside the Solar System are labelled as super-Earths, hot Jupiters and sub-Neptunes, but these planets can be radically different from those of our system, coming in a vast array of forms.

If anything, the discovery of thousands of exoplanets has shown that our Solar System is reassuringly mundane. And the first exoplanet discovered was an example of an object absent from the Solar System. The first planet detected outside of the Solar System was discovered by Aleksander Wolszczan and Dale Frail in January 1992. The duo discovered the rocky exoplanet orbiting the binary PSR B1620-26, consisting of a white dwarf and a pulsar, located over 12,000 light years away.

The following year, a second planet was discovered in the same system – also a terrestrial world. These planets, the two outermost planets of the system, were given the names Poltergeist and Phobetor and represented the first examples of so-called super-Earths. These planets are defined by their masses, which are greater than our planet's but still less than those of the Solar System's ice giants, Uranus and Neptune. The upper limit for the mass of a super-Earth is generally considered to be about ten times that of our planet. But you shouldn't be fooled into thinking that super-Earths bear any other similarities to our planet. The term doesn't say anything about an exoplanet's surface conditions or habitability. As a striking example of this fact, researchers quickly determined that neither Poltergeist nor Phobetor could support life, as they were being blasted by harsh radiation from the pulsar they orbited.

The search for a planet around a star similar to the Sun hit gold in 1995 when Michel Mayor and his then-doctoral student Didier Queloz discovered 51 Pegasi b, or Dimidium, a planet in orbit around a star that resembled our Sun. In October 2019, the Nobel Committee awarded the Nobel Prize in Physics to the duo for their discovery of the planet. But though the star it orbits, 51 Pegasi, is Sun-like, that doesn't mean its planetary system resembles the Solar System. This discovery marked the first detection of a hot Jupiter – a planet with the size and composition of the Solar System's gas giant, but located scorchingly close to its parent star.

"These planets are at an orbital distance closer than Mercury is from the Sun. That means hot Jupiters complete their orbits in only a few days. Due to their location close to their host stars, they're highly irradiated, with temperatures of 1,725 degrees Celsius (3,140 degrees Fahrenheit) or more," Romain Allart, a Trottier postdoctoral fellow at the University of Montréal, Canada, and a team member at the Institute for Research on Exoplanets, tells All About Space.

Not only was 51 Pegasi b an early hint to astronomers that the universe is a wilder and more varied place than they may have

"It's very fundamental to being human to ask the question if there are planets elsewhere"
Nikku Madhusudhan

One goal of exoplanet research is to find a future base for humanity

Present exoplanet missions

GAIA *Launched 2013*
The Gaia mission aims to observe 1 billion astronomical objects. It detected its first confirmed exoplanet in March 2021.

TESS *Launched 2018*
By May 2023, the Transiting Exoplanet Survey Satellite had identified over 6,500 potential exoplanets, of which over 330 were been confirmed so far.

CHEOPS *Launched 2019*
The CHaracterising ExOPlanets Satellite is the first mission dedicated to investigating exoplanet atmospheres.

JWST *Launched 2021*
One of the James Webb Space Telescope's primary missions will be to investigate exoplanet atmospheres.

Kepler-47 system

Exoplanets discovered by Kepler orbiting a binary pair, with one in the Goldilocks zone around the stars

Kepler-47c

Kepler-47B

Solar system

Mercury

Venus

Earth

Mars

Habitable zones

Finding an Earth-like planet

Is there a planet identical to ours somewhere in the Milky Way?

It's perhaps of little wonder that with our own planetary biases, much of our search for exoplanets has hinged on the search for worlds similar to Earth in both composition and location in the habitable zone of their main sequence stars. Another condition that would make a world similar to Earth would be a magnetic field, which can prevent an atmosphere from being stripped away and thus sustain life.

Several of the planets in the exoplanet catalogue are similar to Earth in a range of ways, be it distance from their star, their mass or their radius. Below are 21 candidates for Earth-analogue worlds. But even if these worlds do fit the description of a 'planet B' that could be occupied by humans in the future, crewed travel to worlds outside the Solar System is thus far unfeasible to say the least, and likely will be for several generations. As Institute for Research on Exoplanets scientist Romain Allart points out, that means despite the wealth of exoplanets discovered, protecting our own world is still of paramount importance.

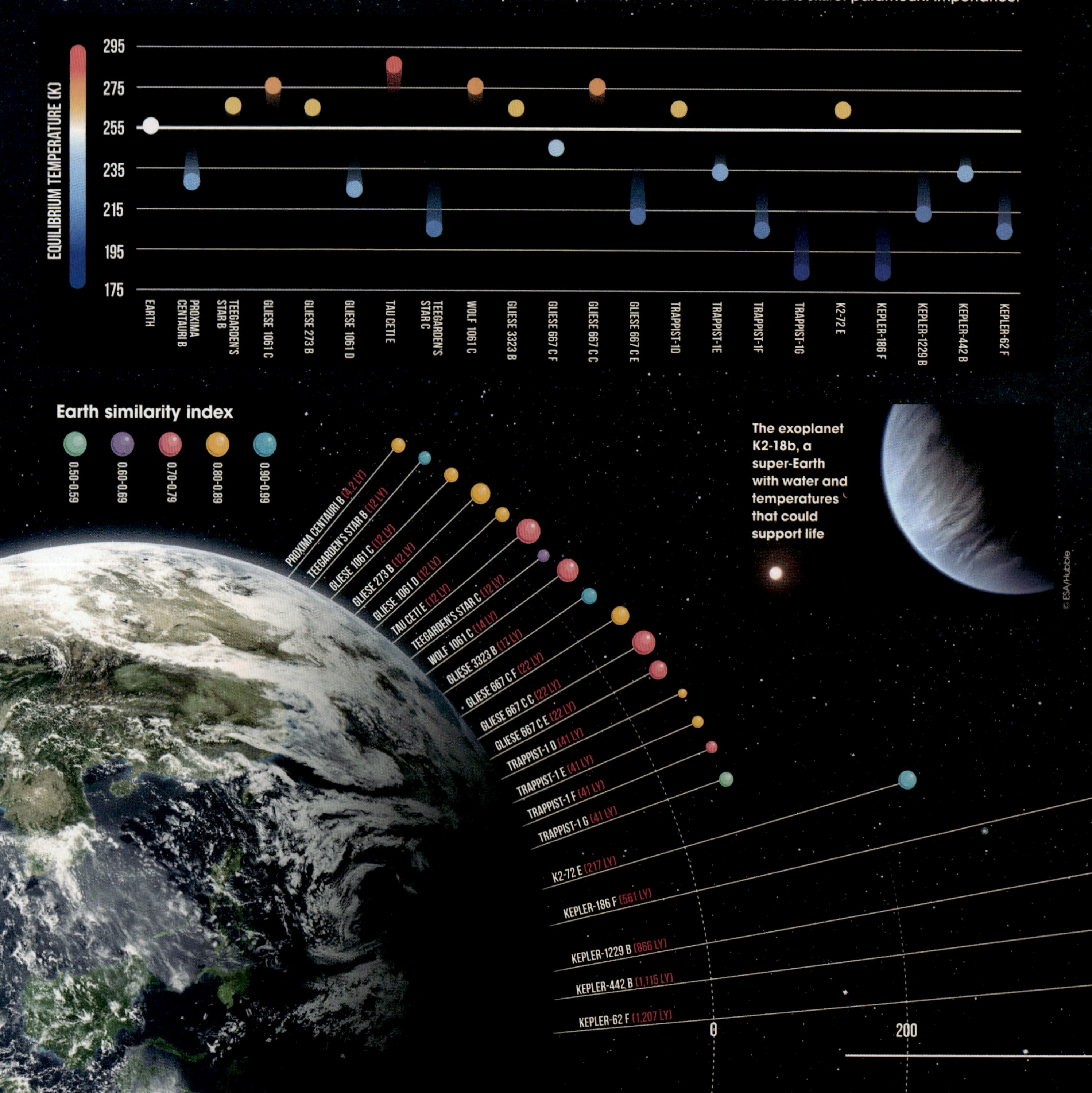

The exoplanet K2-18b, a super-Earth with water and temperatures that could support life

Methods of exoplanet detection

The transit method

The transit method hinges on tiny dips in light from a star caused as a planet crosses its face. Despite only being viable for planets that pass between their star and observers, it's delivered over 3,700 exoplanet discoveries.

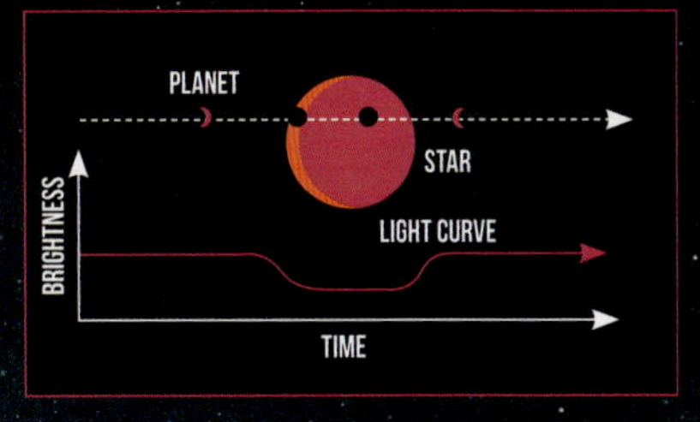

The radial velocity method

A technique based on the fact that a planet and a star orbit a point of mutual mass, meaning the presence of a planet can be seen as a 'wobble' in the star's motion. Also known as the wobble method, this has delivered around 900 discoveries.

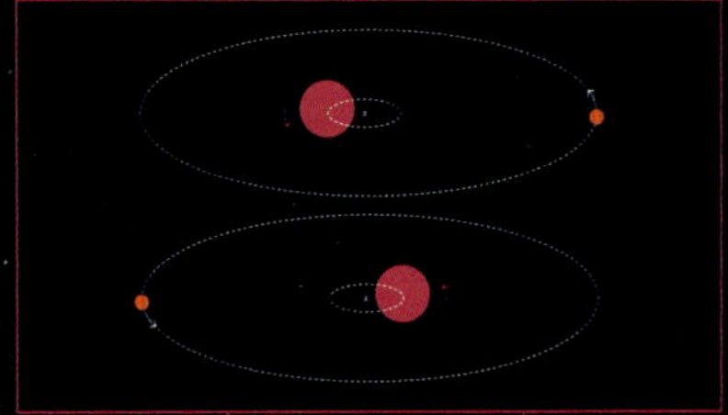

Gravitational microlensing

Responsible for the detection of about 130 planets, lensing occurs when objects of tremendous mass warp the fabric of space. By observing the curvature of light from a distant object, details can be ascertained about intervening objects, like planets.

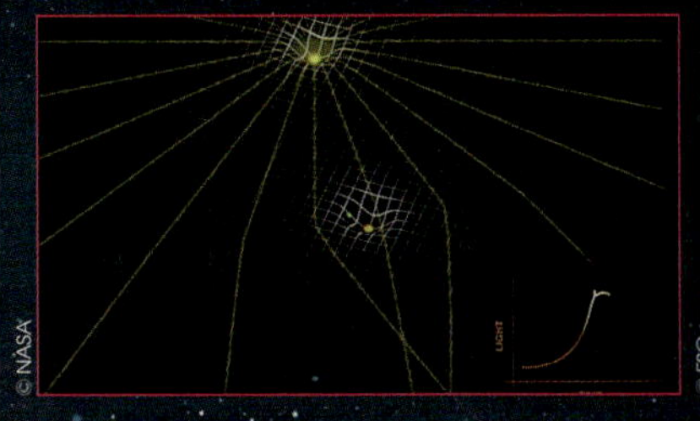

© NASA

Direct imaging

Occasionally, by blocking the light from a parent star astronomers can actually directly image an exoplanet. This method has resulted in the discovery of 54 worlds outside the Solar System. This includes TYC 8998-760-1 b and c, pictured below.

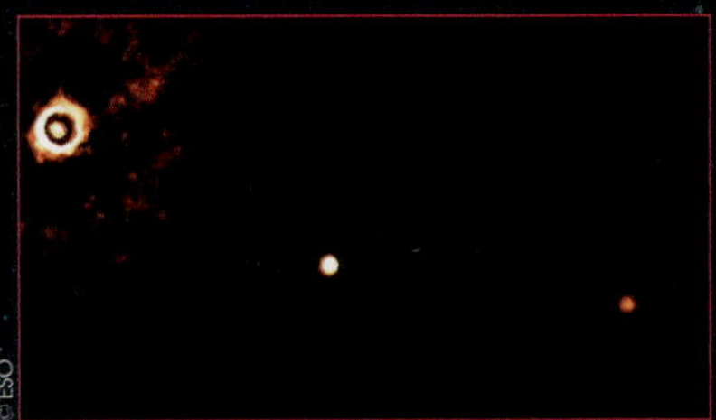

© ESO

The Goldilocks zone

Not too hot. Not too cold, but just right. The habitable zone is defined as the region around a star at which liquid water can exist. In the Solar System, Earth exists within the Goldilocks zone, but so do Mars and Venus. This should suggest that just because a planet exists in a habitable zone doesn't mean it will be abundant with water. Other factors are vital to possessing liquid water, such as planet size and temperature. Exoplanets in the habitable zone could have lost their water as a result of a runaway greenhouse effect like Venus, or through the loss of a magnetic field because they were too small to maintain it, as seems to be the case with Mars. But the term is based on our bias. There's only one planet we are aware of that possesses life, and it sits in this zone. Too hot and water boils; if water exists on worlds too close to their star it only does so as vapour. Too cold and liquid water freezes to ice. This may not totally rule out life, however, as liquid water could exist beneath an icy shell. But just right, and liquid water could exist.

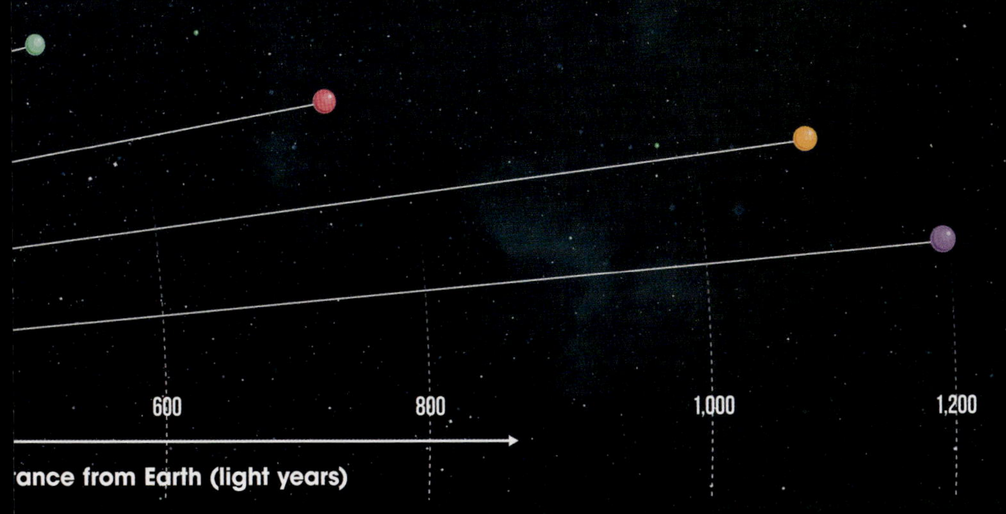

"As soon as exoplanets were discovered, that opened an entirely new window into the universe"
Nikku Madhusudhan

previously suspected when it comes to planets, but hot Jupiters would also become mainstays of the exoplanet catalogue. "Hot Jupiters are actually not so common in the universe, but due to instrumental biases they are extremely common in the current exoplanet catalogue," Allart, who was part of the team that investigated the hot Jupiter WASP-76b, explains. "Because they are close, large and massive, the radial velocity and transit techniques are efficient to detect hot Jupiters; these two techniques have discovered almost all exoplanets up until now."

However, in terms of exoplanet populations, Madhusudhan says that sub-Neptunes – planets with a smaller radius than Neptune but a larger mass, or those with a smaller mass than Neptune but a larger radius – seem to dominate the Milky Way. "The realisation that small planets are extremely common elsewhere is another major milestone," Madhusudhan adds.

One milestone in exoplanet research that is currently ongoing and will develop exponentially in the future, the astrophysicist says, is the investigation of these more diminutive planets' atmospheres and the search for water. An exoplanet transiting the face of its host doesn't just provide a great way for astronomers to spot such a world via the dip in light output from the star it causes. The transit method has also proved a good way of assessing the composition of a planet's atmosphere. This is because atoms and molecules absorb light at characteristic wavelengths. By observing the 'gaps' in the light signatures of stars as they shine through planets' atmospheres, astronomers can see what elements make up these gaseous envelopes.

In 1999, Greg Henry and David Charbonneau used the transit method to detect and observe an exoplanet as it passed in front of the star HD 209458, revealing the planet, HD 209458 b, had an atmosphere of oxygen, nitrogen, carbon and, importantly, water. This atmosphere is being stripped away from the world, leaving a trail behind it that is similar to that of a comet.

Madhusudhan points out that since 1999, and particularly in the past decade, atmospheric observations of exoplanets have taken off in a big way, with the first robust measurements of water vapour in the atmospheres of these planets made. But unfortunately, as was the case with HD 209458 b, many of these detections, which have become commonplace, tell us little about the possibility of life existing there. "Hot, giant planets are where we have detected water, for the most part as water vapour. And there's no scope of life on these planets," the astrophysicist says.

Excitingly, however, this is beginning to change. Madhusudhan, who is the editor of a forthcoming book called Exofrontiers which collects pioneering work from the exoplanet science community, points out that our methods of examining atmospheres have improved to the point where we are now able to detect chemical elements around much smaller planets. This includes Earth-like worlds in the habitable zones of planets where conditions are just right to allow for the existence of liquid water. "We are able to detect small Earth-sized planets in the habitable zones of their host stars around nearby stars. And this is especially true for small stars called M dwarfs," Madhusudhan says, referencing the planets in the TRAPPIST-1 system in particular.

Discovered in 2017 and located just 40 million light years away, the system contains at least seven rocky terrestrial worlds, all of which exist at a suitable distance from their red dwarf to facilitate the existence of water on their surface. "These are all small, rocky, Earth-like planets at the right distances for habitability around their host stars," Madhusudhan says. Observations of the TRAPPIST-1 planets conducted in February 2018 revealed that some of them may even be able to harbour more liquid water and wider oceans than Earth. This makes the system one of the prime targets for atmospheric investigations by future telescopes, including the James Webb Space Telescope (JWST) currently on its way to orbit.

This life-searching, atmosphere-investigating aspect wasn't part of the JWST's mission when the plans for a ten-metre (32.8-foot) passively cooled near-infrared telescope in a high-Earth orbit were initially floated in 1989. In the last year of the 1980s, astronomers hadn't even

Future exoplanet missions

PLATO *Planned launch 2026*
The PLanetary Transits and Oscillations of stars telescope will search for an categorise planets in their stars' habitable zones to search for other Earth-like worlds.

Nancy Grace Roman Space Telescope *Planned launch 2026-27*
The NGRST's objectives include completing a census of thousands of exoplanets, and performing the first direct imaging of nearby exoplanets.

ARIEL *Planned launch 2029*
The Atmospheric Remote-sensing Infrared Exoplanet Large-survey will aim to observe, study and characterise at least 1,000 known exoplanets.

"Exoplanet diversity is so rich that even the best sci-fi authors couldn't have imagined it"
Romain Allart

WASP-76b has such a blistering temperature that molten iron rains down on the planet's cooler side

Hot Jupiters come in a wide variety of shapes and sizes and are easy to detect because they transit their stars often

discovered planets around other stars, and the Hubble Space Telescope, which would make an important contribution to this search, was still a year from launch.

Various teams of astronomers are currently applying for observation time with the new space telescope so they can investigate planets outside the Solar System. This includes Madhusudhan, who will be leading a team working with the JWST to investigate exoplanet atmospheres in unprecedented detail.

"We are indeed in the golden age of exoplanet science, but we are also on the verge of a major revolution in modern astronomy," the astrophysicist says. And while even Webb still won't be able to conclusively tell if a planet is hosting life, its observing power brings humanity tantalisingly close to the detection of molecules that hint at the presence of living organisms. This will lay further groundwork for future missions. "We are the fortunate generation that might witness the discovery of life elsewhere," Madhusudhan says. "We have been dreaming of that for thousands of years, and we happen to be that blink-of-an-eye generation in which that momentous discovery is going to happen. To me that is huge."

Not only this, but Madhusudhan is researching so-called Hycean worlds: water-rich planets with surfaces covered almost entirely in oceans and with atmospheres made up of mostly molecular hydrogen. These hypothetical worlds could potentially redefine the limits of what we consider the habitable zone. This gives researchers targets outside the traditional habitable zone to include in the search for the telltale signatures of life.

And nothing says 'casting a wider net' like the revelation in 2021 that astronomers may have caught a hint of the first exoplanet ever to be detected outside of the Milky Way. The team, including Nia Imara from the University of California, may have detected a Saturn-sized exoplanet 28 million light years from Earth in the galaxy Messier 51. This extragalactic exoplanet seems to be orbiting a high-mass compact object such as a neutron star or a black hole.

"Surprisingly, we are only scratching the surface, as we now think that almost one star in every two hosts a planet, and there are hundreds of billions of stars in our galaxy alone, and there are billions of galaxies in the universe," Allart says. "Exoplanet diversity is already so rich that even the best sci-fi authors couldn't have imagined it. It's amazing to discover more and more strange exoplanet systems and worlds."

Allart continues by explaining that despite this wealth of planets, and our increasing knowledge about them, protecting our own world is still of paramount importance. "The Solar System, and in particular Earth, remains unique in the diversity of exoplanets. Therefore it's important to understand that there's no planet B," he concludes.

Robert Lea
Space science writer
Rob is a science writer with a degree in physics and astronomy. He specialises in physics, astronomy, astrophysics and quantum physics.

How many stars are there in the universe?

With a fleet of missions scouring the cosmos, will it ever be possible to come to an estimate?

© Getty

Looking up into the night sky, you might wonder just how many stars are in the universe. It's challenging enough for an amateur astronomer to count the number of naked-eye stars that are visible, and with bigger telescopes, more stars come into view, making counting them a lengthy process. So how do astronomers figure out how many stars are in the universe?

The first tricky part is trying to define what 'universe' means, says David Kornreich, a professor at Ithaca College in New York State. He was also the founder of the 'Ask An Astronomer' service at Cornell University. "I don't know, because I don't know if the universe is infinitely large or not," he said. The observable universe appears to go back in time by about 13.8 billion years, but beyond what we could see there could be much more. Some astronomers also think that we may live in a 'multiverse', where there would be other universes like ours contained in some sort of larger entity.

The simplest answer may be to estimate the number of stars in a typical galaxy, and then multiply that by the estimated number of galaxies in the universe. But even that is difficult, as some galaxies shine better in visible light or in infrared. There are also estimation hurdles that must be overcome. In October 2016, deep-field images from Hubble suggested that there are about 2 trillion galaxies in the observable universe, or about ten times more than previously suggested. Speaking with All About Space, the lead author of a new study published in Nature, Christopher Conselice, a professor of astrophysics at the University of Manchester in the UK, says there are about 100 million stars in the average galaxy.

Even telescopes may not be able to view all the stars in a galaxy, however. A 2008 estimate by the Sloan Digital Sky Survey – which catalogues all the observable objects in a third of the sky – found about 48 million stars, roughly half of what astronomers expected to see. A star like our own Sun may not even show up in such a catalogue, so many astronomers estimate the number of stars in a galaxy based on its mass. This has its own difficulties, since dark matter and galactic rotation must be filtered out before making an estimate.

Missions such as Gaia, a European Space Agency (ESA) probe that launched in 2013, may

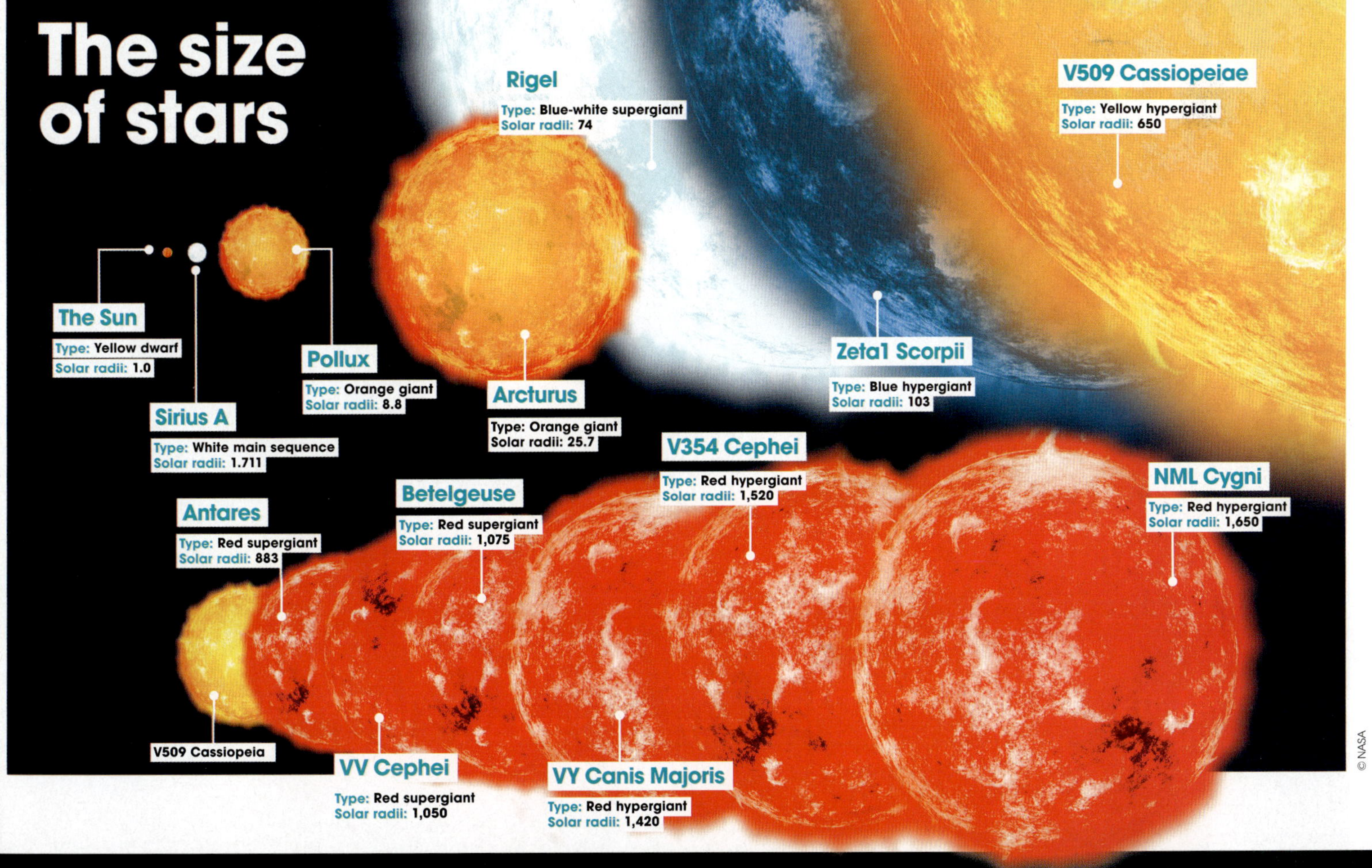

The Milky Way mapper

How Gaia has been constructed to create a 3D map of the Milky Way

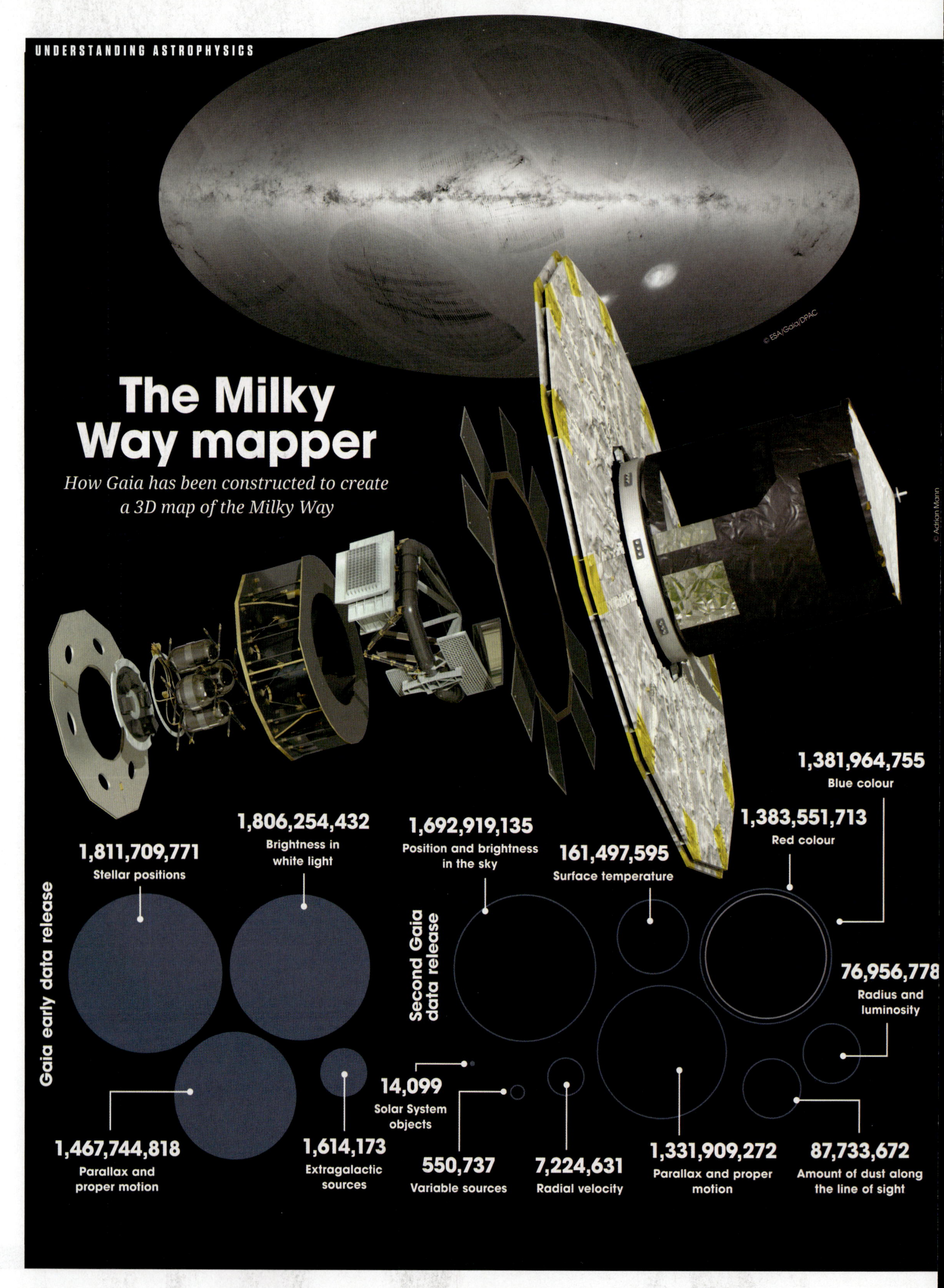

1 Not a scorcher
The sunshield keeps the scientific instruments at a constant temperature of -110 degrees Celsius (-166 degrees Fahrenheit). It has thermal insulation and is resistant to temperatures between -170 and 70 degrees Celsius (-274 and 158 degrees Fahrenheit).

2 Sensitive sensors
Gaia has sensors that are so powerful they can detect faint objects with a luminosity 400,000 times lower than the naked eye is capable of seeing.

3 Strong signal
The average UK broadband speed is 14.7Mbps. While Gaia's 8.7Mbps may seem slow in comparison, considering it's beaming from 1.5 million kilometres (932,000 miles) away, it's quite good.

4 Packing light?
Gaia's got some payload: ten mirrors, an astrometry function, a photometry function and a spectrometry function all within the one unit on a spacecraft three metres (9.8 feet) high.

5 Snap happy
Chinese manufacturer Oppo may be creating a 50-megapixel smartphone camera, but Gaia boasts a photometer with a resolution of 1,000 megapixels. That's sharp.

6 Data hungry
So much data will be taken by Gaia that if you were to put everything onto a DVD, you would need 200,000 discs: that's more than 100 terabytes worth.

7 Silicon carbide
The material that goes into making cutting tools and sandpaper is silicon carbide. It's tough stuff – a mix of pure silica sand and carbon – and it makes up Gaia's structure.

8 Huge sunshield
A cutting-edge sunshield made from carbon-fibre-reinforced composite protects the Gaia spacecraft during its mission. It's like a big skirt at ten metres (33 feet) in diameter.

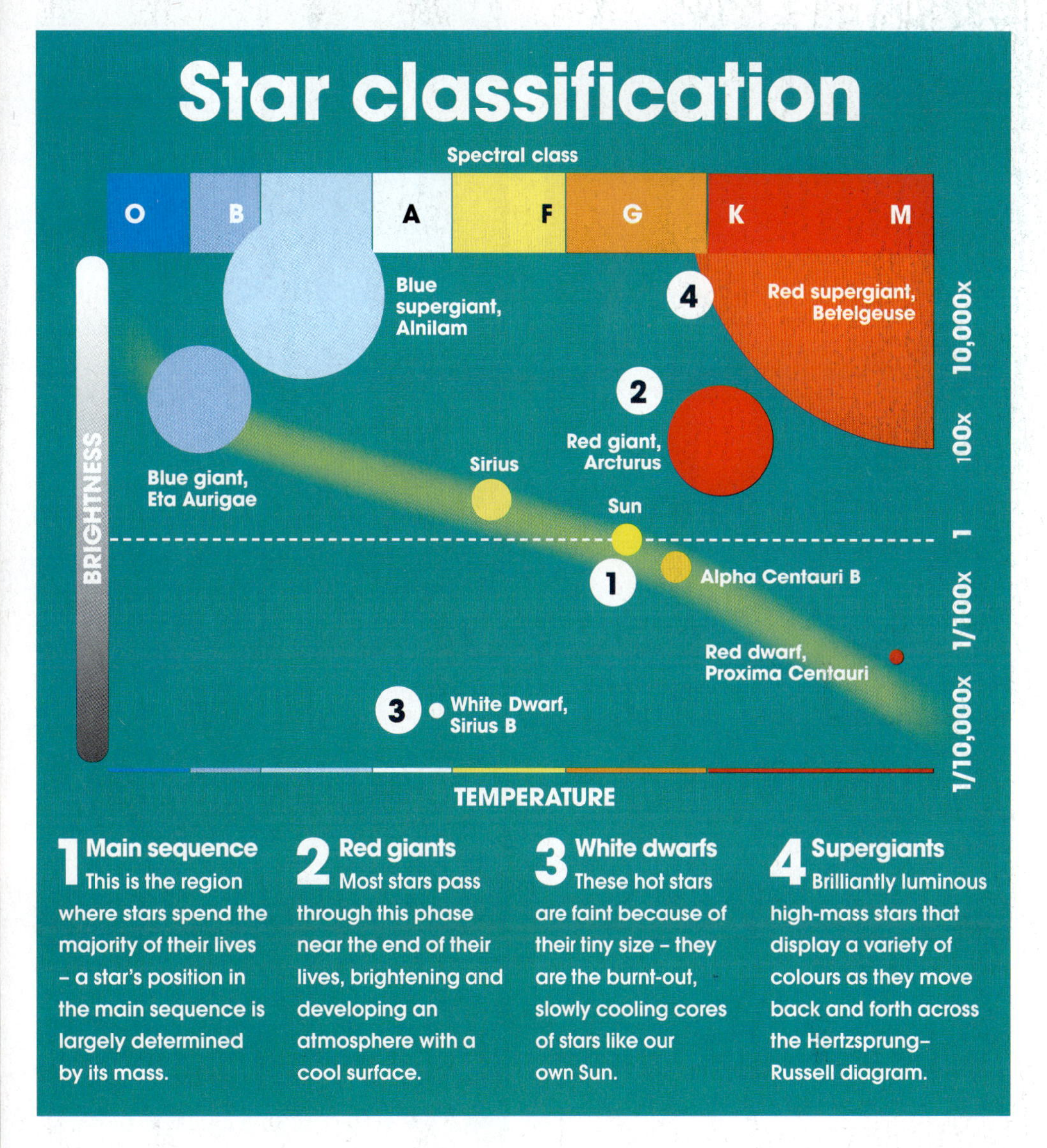
Star classification
Spectral class
O
B
A
F
G
K
M
Blue supergiant, Alnilam
4
Red supergiant, Betelgeuse
2
Red giant, Arcturus
Blue giant, Eta Aurigae
Sirius
Sun
1
Alpha Centauri B
Red dwarf, Proxima Centauri
3
White Dwarf, Sirius B
BRIGHTNESS
TEMPERATURE
10,000x
100x
1
1/100x
1/10,000x
1 Main sequence
This is the region where stars spend the majority of their lives – a star's position in the main sequence is largely determined by its mass.
2 Red giants
Most stars pass through this phase near the end of their lives, brightening and developing an atmosphere with a cool surface.
3 White dwarfs
These hot stars are faint because of their tiny size – they are the burnt-out, slowly cooling cores of stars like our own Sun.
4 Supergiants
Brilliantly luminous high-mass stars that display a variety of colours as they move back and forth across the Hertzsprung-Russell diagram.

provide further answers. Gaia aims to precisely map about 1 billion stars in the Milky Way. It builds on the previous Hipparcos mission, which precisely located 100,000 stars and also mapped 1 million stars to a lesser precision.

"Gaia will monitor each of its 1 billion target stars 70 times during a five-year period, precisely charting their positions, distances, movements and changes in brightness," the ESA states on its website. "Combined, these measurements will build an unprecedented picture of the structure and evolution of our galaxy. Thanks to missions like these, we are one step closer to providing a more reliable estimate to a question asked so often: 'How many stars are there in the universe?'"

Even if we narrow down the definition to the 'observable' universe – just what we can see – estimating the number of stars within it requires knowing just how big the universe is. The first complication is that the universe itself is expanding, and the second complication is that space-time can curve.

To take a simple example, light from the objects farthest away from us would take approximately 13.8 billion years to travel to Earth, taking into account that the very youngest objects would be shrouded because light couldn't carry in the early universe. The radius of the observable universe should be 13.8 billion light years, since light only has that long to reach us.

Or should it? "It's a logical way to define distance, but not how a relativist defines distance," Kornreich says. A relativist would use a device such as a metre stick, measuring the distance along that device and then extending it as long as needed. This produces a different answer, and some sources define the universe as being 48 billion light years

The darker the sky, the more stars become visible

in radius. But sources vary on this number because space-time can curve. As the observer does the measurement with the metre stick, light travels at the same time and influences the measurements.

It's easier to count stars when they are inside galaxies, since that's where they tend to cluster. But to even begin to estimate the number of stars in the universe, you would need to estimate the number of galaxies and come up with some sort of an average, even though galaxies vary wildly.

Some estimates peg the Milky Way's star mass at around 100 billion solar masses, or 100 billion times the mass of the Sun. Averaging out the types of stars within our galaxy, this would produce an answer of about 100 billion stars in the galaxy. This is subject to change, however, depending on how many stars are bigger and smaller than our Sun. Other estimates say the Milky Way could have 200 billion stars or more.

The amount of galaxies that permeate the universe is an astonishing number, as shown by some imaging experiments performed by the Hubble Space Telescope. Several times over the years, Hubble has pointed a detector at a tiny spot in the sky to count galaxies, performing the work again after the telescope was upgraded by astronauts during the Space Shuttle era.

A 1995 exposure of a small spot in Ursa Major revealed about 3,000 faint galaxies. In 2003 to 2004, utilising Hubble's upgraded instruments, scientists looked at a smaller spot in the southern constellation of Fornax and found 10,000 galaxies. An even more detailed investigation in Fornax in 2012, with even better instruments, showed about 5,500 galaxies.

Kornreich uses a very rough estimate of 10 trillion galaxies in the universe. Multiplying that by the Milky Way's estimated 100 billion stars results in a large number indeed: 124 stars, or a one with 24 zeros after it. That's 1 septillion in the American numbering system or 1 quadrillion in the European system. Kornreich emphasises that number is likely a gross underestimation, as more detailed studies of the universe will reveal even more galaxies.

A star is born

Compared to other stars, the Sun is in the middle of the pack when it comes to size and temperature

1 Almost a star

A protostar is a ball-shaped mass in the early stages of becoming a star. It's irregularly shaped and contains dust as well as gas, formed during the collapse of a giant molecular cloud. The protostar stage in a star's life cycle can last for a 100,000 years as it continues to heat up and become denser.

Red dwarf
3
Low-mass stars
1
Protostars
Sun-like stars
Giant molecular cloud

2 Star or planet?

Brown dwarfs are sometimes not even considered stars at all, but instead sub-stellar bodies. They are incredibly small in relation to other types of stars, and never attained high enough temperatures, masses or enough pressure at the core for nuclear fusion to actually occur. They are below the main sequence on the Hertzsprung-Russell diagram. They have a radius about the size of Jupiter, and are sometimes difficult to distinguish from gaseous planets because of their size and composition of helium and hydrogen.

Brown dwarfs

Ailsa Harvey
Science and technology journalist
Ailsa is a staff writer for *How It Works* magazine, where she focuses on writing about science, technology, history and the environment.

3 The cool star

Red dwarfs are small and relatively cool, tending to have masses less than one-half that of our Sun. The heat generated by a red dwarf occurs at a slow rate through the nuclear fusion of hydrogen into helium before being transported via convection to its surface.

4 A dead star

White dwarfs are the end of a small star's life cycle. A white dwarf is small, with a volume comparable to that of Earth's, but incredibly dense, with a mass about that of the Sun's. With no energy left, a white dwarf is dim and cool in comparison to larger stars.

5 Stellar remnants

Black dwarfs are the hypothetical next stage of star degeneration, when they become sufficiently cool to no longer emit any heat or light. Because the time required for a white dwarf to reach this state is postulated to be longer than the current age of the universe, none are expected to exist yet. If one were to exist it would be difficult to locate and image due to the lack of emitted radiation.

6 Neutron stars

Neutron stars are a potential next stage in the life cycle of a star. If the mass that remains after a supernova is up to three times that of the Sun, it becomes a neutron star. This means that the star only consists of neutrons, particles that don't carry an electrical charge.

7 The rarest stars

Supergiants are among the rarest stars, and can be as large as our entire Solar System. Supergiants can also be tens of thousands of times brighter than the Sun and have radii of up to a thousand times that of the Sun.

8 The absence of light

Stellar black holes are thought to be the end of the life cycle for supergiant stars with masses more than three times that of our Sun. After going supernova, some of these stars leave remnants so heavy that they continue to remain gravitationally unstable.

The edge of the universe

Finding and defining exomoons

As our catalogue of exoplanets grows, the hunt is on for moons around distant worlds

Astronomers may have finally discovered a moon orbiting around a planet outside our Solar System. The development could have important implications for our understanding of how planetary systems evolve, as well as indicating how typical the planets of our Solar System are. Since the first discovery of planets around stars other than the Sun was made in the early 1990s, our catalogue of planets outside the Solar System has burgeoned, now containing over 4,000 confirmed worlds. One of the most important aspects of our investigation of these extrasolar planets – or exoplanets – is the assessment of how similar or diverse they are in comparison to the planets of the Solar System.

It's unsurprising given how Earth's own Moon has dominated our imagination and played a vital role in both astronomy and humanity's exploration of space that the hunt for moons around these extrasolar worlds – or exomoons – is a subject of intense interest. This is compounded by the fact that moons are abundant in the Solar System, with an average of 20 moons for each planet. And we have no reason to believe that our planetary system is unique in this respect, making the elusive nature of exomoons confusing and frustrating.

To understand why the discovery of exomoons may be important, it's worth considering just how influential some researchers believe our natural satellite was to the development of life on Earth. This is coupled with the fact that many scientists theorise that if life exists elsewhere in the Solar System, aside from Mars, the most likely places to find it would be on the moons of gas giants Jupiter and Saturn, with the icy moon Enceladus being a particularly enticing suspect. This means that exomoons could be a target for habitability investigations in their own right, especially when they orbit gas giants and could potentially be the size of Earth, or even larger.

Determining if an exoplanet is orbited by a moon could be important for investigations of that planet's habitability potential in another way. For rocky terrestrial exoplanets in the habitable zones of their stars – the area that has the right temperature to allow for the existence of liquid water – an undetected moon may add an additional source of thermal or reflected light that hinders the determination of characteristics belonging to the planet itself.

Observations of our Solar System, in which moons are plentiful, suggest that such objects should be extremely common throughout other systems. However, until recently these exomoons have proven to be particularly elusive. This could be because our detection

Types of exomoon

Snowball exomoon

A snowball exomoon is one where the entire moon is permanently frozen, and less than one-tenth is habitable. These worlds are likely to have been born far from their stars, or were perhaps moved there from a position nearer the star.

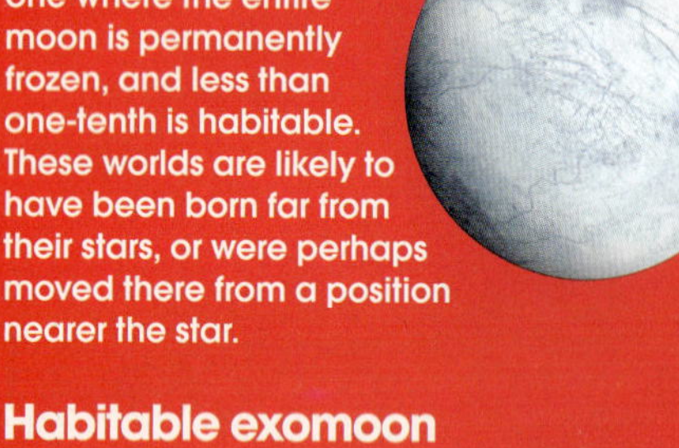

Habitable exomoon

Habitable exomoons are those where at least a tenth of the surface could support liquid water. These will need to be in the star's habitable zone, as well as the right distance from the planet, but could be great bets for finding life elsewhere.

Transient exomoon

A transient exomoon is said to be sort of habitable, but its habitability changes dramatically over time. The idea that a moon may shift in and out of habitability will need careful consideration when we look for and study exomoons.

Hot exomoon

Researchers Duncan Forgan and Vergil Yotov came up with a system for classifying exomoons in 2014. One class was a hot exomoon, defined as one with an average surface temperature of more than 100 degrees Celsius (210 degrees Fahrenheit), with less than one-tenth being habitable.

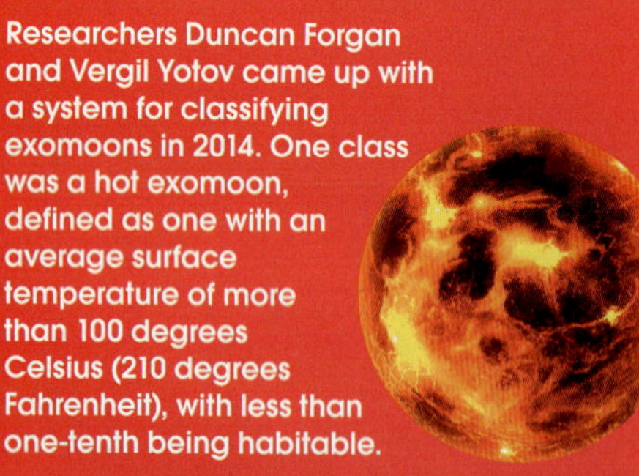

If alien worlds are anything like those in our Solar System, moons should be plentiful

Zodiacal light over La Silla Observatory in Chile

methods are not sensitive enough as of yet to spot these moons outside the Solar System, or because moons are less common in other planetary systems for some reason. Determining which of these possibilities is correct is an important and missing aspect of our ongoing investigations of extrasolar planetary systems.

Discovering exoplanets, though not easy, relies on several tried-and-tested methods. The most powerful of these is the transit technique – used by both Kepler and NASA's Transiting Exoplanet Survey Satellite (TESS) – which hinges on observing the tiny dips in light from a distant star as an orbiting planet crosses its face. The transit technique used to detect exoplanets is also being used to hunt for exomoons, though undeniably with less success. To understand why the transit technique – the most successful exoplanet-hunting method, yielding thousand of discoveries – may not be as successful in detecting exomoons, consider what astronomers must spot to indicate the presence of a lunar body.

Like cosmic nesting dolls, to spot an exomoon using the transit method, astronomers would need to observe a transit within a transit – a situation in which as a planet transits its star, its moon is also visible transiting the star's face. Even if this occurred and we were lucky enough to catch it, the observation would have to be made with equipment sensitive enough to spot the exomoon and distinguish the dip in light it causes from the dip in light caused by its parent planet.

Furthermore, even large exoplanets cause just tiny dips in the light signature of their parent stars, so imagine how small a reduction in light would be when caused by an exomoon, being much more diminutive than its host planet. Another issue with detecting exomoons is related to the kinds of exoplanets that the highly successful transit method is best at spotting, as these may not be planets that are best suited to hosting moons. The transit method excels at discovering large planets that are very close to their parent stars. This is based on the simple logic that the larger a planet is, and the closer it exists to its star, the more light it will block when it crosses that star's face. This creates a more prominent signal for astronomers to identify and distinguish from background noise.

Extremely proximate planets are often tidally locked to their stars – as is the case with Mercury – and thus their rotation slows. This proximity can mean that one face of the planet permanently faces its star, becoming blisteringly hot, resulting in a world with a permanent dayside and nightside. But it has another effect, and it's one that has a significant impact on the chance of an exoplanet holding onto natural satellites. When orbiting around a planet, a moon or any other natural satellite will often have a synchronous orbit. For an orbit to be stable, this needs to be within what is called the Hill sphere – the hypothetical region around a planet in which it dominates the gravitational attraction of objects. For exomoons, this is a limiting radius at which they may be gravitationally bound and remain stable.

As the rotation of a planet is slowed, the synchronous orbit is pushed out, away from that world. For tidally locked exoplanets that barely spin – if at all – this orbit is pushed beyond the Hill sphere. This means that any potential moons within the Hill sphere will not have a stable synchronous orbit and will thus spiral towards the planet. Meanwhile, moons outside the Hill sphere will likely escape the planet due to the lack of gravitational influence they experience, thus never reaching the synchronous orbit.

There is another hindrance to exomoon possession experienced by exoplanets that exist close to their parent stars. Many of these exoplanets orbit their stars with a period of a few days, while others take just hours to circle their parent star. In simulations published in May 2021, researchers from the University of Groningen used a simplified model of one exoplanet, orbiting a single star and possessing a single exomoon, to determine how likely such a world would be to hold its moon as different parameters were altered.

Watching the model evolve over a period of 10 billion years, the team tested how changing

Hunting for exomoons

Kepler helped us find many exoplanets that may host natural alien satellites

1 Sunshade
As the name suggests, the sunshade blocked the Sun's rays from the photometer so it could observe the universe without being blinded or obstructed.

2 Photometer
The most important instrument on Kepler was the photometer, which looked at stars to notice dips in their light as planets pass in front, known as a transit.

3 Reaction wheels
These were used to point the telescope at distant stars in order to find planets.

4 Solar array
Kepler had to roll 90 degrees every three months to keep its solar panels pointed at the Sun and keep the telescope powered.

5 High-gain antenna
This was used to communicate with Earth, receive commands and send back data on any exoplanets Kepler found.

Solar balance ridge

Photons in light emitted by the Sun exerted pressure on the Kepler space telescope, allowing the spacecraft to be properly balanced against the pressure.

Start
Spacecraft rotated to prevent sunlight from entering the telescope.

End

Field of view 2

Mission continues

Sun

Photons

Start
Solar panel illuminated.

Field of view 1

Top-down views of the spacecraft

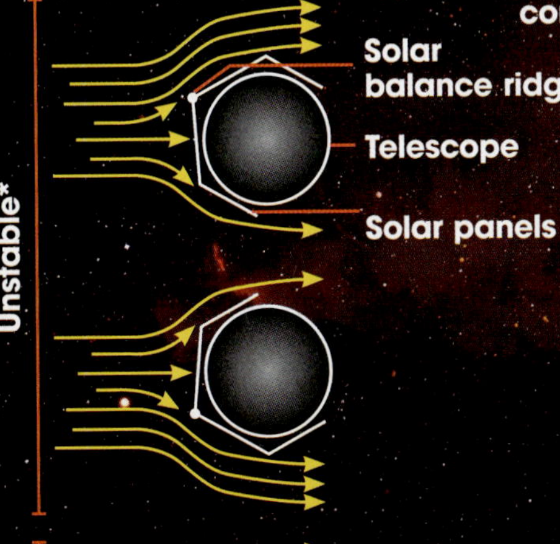

When the spacecraft was balanced, it was stable enough to monitor distant worlds that could harbour exomoons. Kepler usually measured the same area of sky for around 83 days, until it was necessary to rotate it to stop sunlight from entering the telescope.

* Unbalanced solar pressure causes spacecraft to roll

the masses of the components of the system changed the prospects of the planet holding onto its moon. They also adjusted the distance at which the exoplanet orbits its star. This highly simplified model suggested that the orbital period of the exoplanet was key to retaining an exomoon. For exoplanets with orbits ranging from 300 days – similar to the orbit of Earth around the Sun – to ten days, the survival rate of exomoons dropped from 70 down to 0 per cent. The team's model suggests that for planets so close to their host stars that they complete an orbit in less than ten days, there was no stable orbit for a moon.

These results reflect observations of our own planetary system, with the two innermost planets – Mercury and Venus – lacking moons and Earth possessing its own large natural satellite. Jupiter and Saturn, both larger planets that are more distant from the Sun, host large collections of moons. Because the majority of the planets in our current exoplanet catalogue were discovered using the transit method, and this method favours planets close to their parent stars, there's a bias in the exoplanet catalogue that disfavours planets with exomoons. Yet despite the difficulty associated with detecting exomoons, several possible identifications have been made. But these potential exomoon detections are still controversial.

In January 2022, a new paper published in the journal Nature Astronomy revealed that astronomers had discovered a signal that indicates the presence of a moon outside the Solar System. The discovery was made during an investigation of observational data of 70 suspected massive and cool exoplanets that cross – or transit – the face of their parent stars. These exoplanet candidates had been discovered by the Kepler space telescope, which launched in 2009 and was retired in 2018. In this survey, the research team discovered just one signal that could indicate the presence of an exomoon, which has been named Kepler 1708b-i.

The potential exomoon has a proposed radius around 2.6 times that of Earth's. It orbits a world about the size of Jupiter that takes approximately 400 days to complete an orbit of its star. Whether scientists come to consider Kepler 1708b-i to be the first detected exomoon hinges on further investigation of a signal determined by the same group of astronomers in 2018. It's an observation that has been the subject of intense debate in scientific circles.

In a separate paper published in The Astrophysical Journal, a team of astronomers revealed a signal that could indicate the presence of an exomoon around the exoplanet Kepler-1625b, a Jupiter-sized gas giant orbiting a yellow star over 8,000 light years away from Earth. The team spotted the telltale dip in light caused as an object crosses the face of a star, suggesting that it was possibly caused by an exomoon, which was given the designation Kepler-1625b I. The observations made by the team – collected with both the Kepler and Hubble space telescopes – were reassessed the following year by two separate teams of researchers. Both teams were sceptical that an

How to find an alien moon

The different methods being used to hunt for exomoons

Transit timing method

The transit timing method works by measuring a change in the regular transit of a planet. As a planet swings around a star, if there are other objects in its vicinity then it can affect how often a transit is completed. This often indicates the presence of other planets in the system, but in theory the method could be used to hunt for large exomoons as well, revealing their tug on their parent planets. It's hard to distinguish between an exoplanet and an exomoon via this method, though, so it might be difficult to make a discovery with it.

Radial velocity method

When an exoplanet orbits a star, its gravity can cause a tug on the star, making its presence noticeable to us on Earth. The star's spectrum of light is shifted to the blue end of the spectrum if the planet's coming towards us, and the red end if it's going away. Using this method, some characteristics of the planet can be determined. Like the transit timing method, it's thought that the same principle could be applied to exomoons, noticing their effect on a planet. Again, though, the effect is so small that it would be hard to find exomoons using this method.

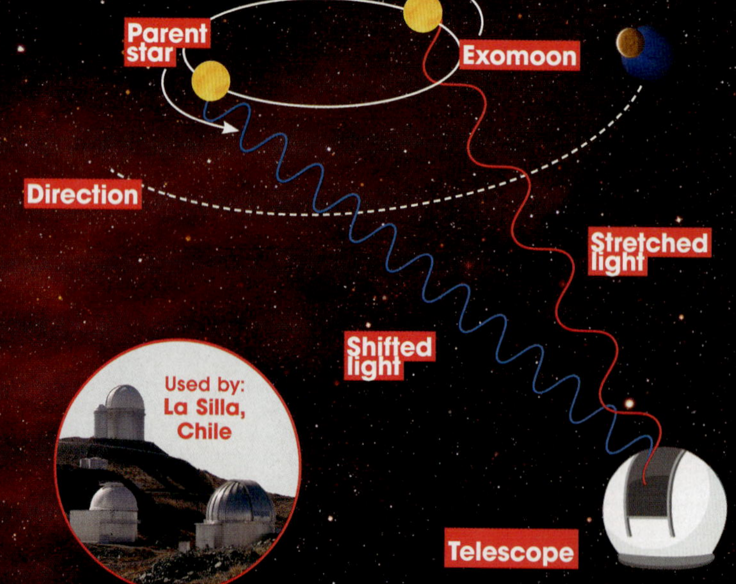

exomoon had been detected around Kepler-1625b, albeit to varying degrees.

In a paper published to arxiv.org on 25 April 2019, one research team, including Laura Kreidberg from the Harvard-Smithsonian Center for Astrophysics, failed to see a moonlike dip in the initial data. They concluded that the light curve of the star Kepler-1625 best conformed to a star transited by a planet alone. A second independent team, including René Heller from the Max Planck Institute for Solar System Research, was only slightly more optimistic. They analysed the same data, finding that signs of Kepler-1625b I were inconsistent.

In a paper published in Astronomy & Astrophysics, Heller and his colleagues point to unknown systematic errors in the Kepler and Hubble data making the finding unreliable, writing that: "Although we find a similar solution to the planet-moon model to that previously proposed, careful consideration of its statistical evidence leads us to believe that this is not a secure exomoon detection."

With so many stars in our galaxy alone, there surely must be an exomoon out there

Transit method

Many exoplanets have been found using the transit method. As a planet periodically orbits its star, it causes a dip in the star's light – or flux – if it crosses our line of sight. Measuring this dip, we can then determine the orbital period and even the size of the planet. The same method could also be used for exomoons, noticing the combined dip produced by both the moon and its planet. Indeed, as the moon will be moving around the planet, it may block a different amount of light depending on its position, giving us a key insight.

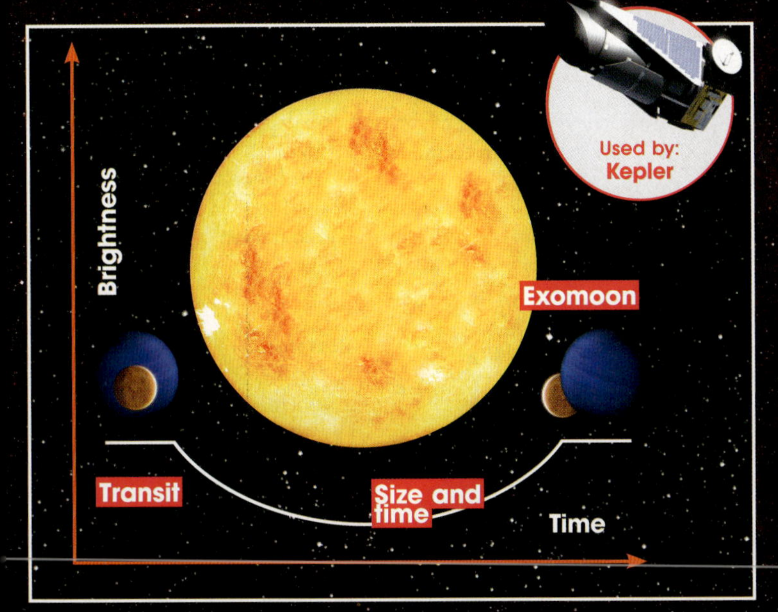

Microlensing

Gravitational microlensing is essentially the result of an object in the universe magnifying a much more distant object. For example, an intermediate star can boost the light of a more distant galaxy, although any given event will occur only once. If the intermediate star has a planet orbiting it, then the effect is altered, and the presence of the planet can be deduced. More recently, it has been suggested that this same method could be used to detect exomoons.

Used by: Keck Observatory, Hawaii

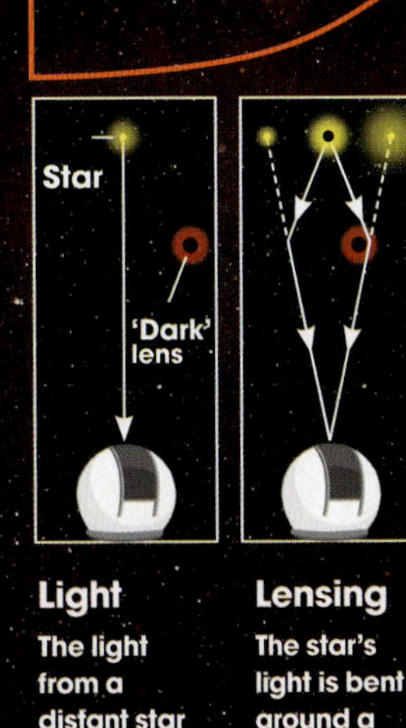

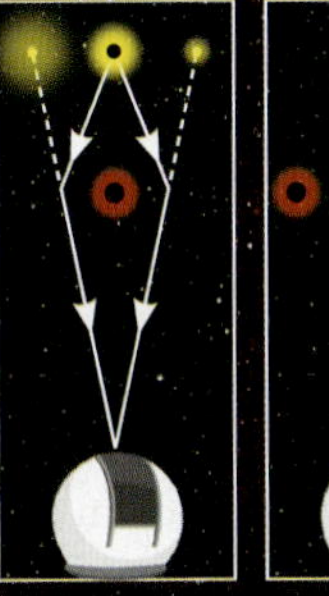

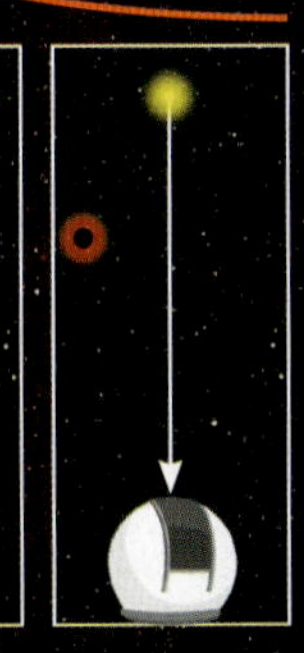

Light The light from a distant star travels and is picked up by telescopes.

Lensing The star's light is bent around a massive foreground object.

Doubled The effect causes us to see multiple versions of the distant star.

Ring The effect can cause a ring of light known as an Einstein ring.

Unique These events only happen once, so there's no opportunity for a repeat.

With more and more exoplanets being discovered, a picture has developed of a universe that is far more varied in terms of worlds than we previously believed. Astronomers have discovered examples of planets for which there are no analogues in our increasingly mundane Solar System. This includes exoplanets like hot Jupiters, those larger than the Solar System's most massive planet which exist so close to their stars they whip around in days or even hours, and rogue planets that wander the galaxy alone, separated from their parent stars. There's a chance that as we discover more exomoons this diversity will be continued through to these smaller bodies, further revealing that the universe can be a strange place.

One exomoon candidate suggested a few years ago is an example of this, and it could provide a solution to a long-standing astronomical mystery. For some time now astronomers have observed the dimming of a distant star called KIC 8462852, or Tabby's Star as it's more commonly known. Many possible mechanisms have been suggested for this often abrupt dimming, ranging from the relatively mundane, like obscuration by dust, to the more 'out there' like black holes, obscuring comets and even to the seemingly fanciful notion of alien 'megastructures'.

In a 2019 paper published in the Monthly Notices of the Royal Astronomical Society, it was suggested that the dimming could be the result of a melting exomoon. The authors of the study, including University of Columbia astrophysicist Brian Metzger, posit that this exomoon, composed mainly of ice and rock, is evaporating, and in the process is spewing off material that obscures the light from the star. Not only would this explain the gradual dimming of Tabby's Star that saw it drop in brightness by about 14 per cent between 1890 and 1989, it could also explain random or sporadic fluctuations in the star's brightness of between 1 and 22 per cent. The team also considered that this exomoon may have been detached from its exoplanet host by powerful tidal interactions with the parent star. This interaction, which could have seen the exoplanet destroyed, may have placed the moon in orbit around Tabby's Star, where it began to slowly disintegrate.

Another extreme exomoon candidate spotted in 2019 could be volcanically active like Jupiter's moon Io. Located in the exoplanet system of WASP-49b, astronomers discovered a signal

Illustration of a rocky world with two exomoons in orbit

Exomoon candidates

There are a few examples that need further study for confirmation

Kepler-1708b I

Distance from Earth: 5,436 light years

Host planet: Jupiter-sized planet Kepler 1708b

Host star: F-type star Kepler 1708

Exomoon size: Mini-Neptune

Discovered: 2022

Kepler-1625b I

Distance from Earth: 8,154 light years

Host planet: Jupiter-sized gas giant Kepler-1625b

Host star: Kepler-1625 has a similar mass to the Sun but is 70 per cent larger

Discovered: 2018

Size: Neptune-sized

WASP-49b's 'Exo-Io'

Distance from Earth: 635 light years

Host planet: WASP-49b a gas giant with a mass 0.37 times that of Jupiter and a similar radius

Host star: WASP-49, a yellow dwarf main-sequence star

Size: Unknown

Discovered: 2022

Why our Moon is important

From protecting us from asteroids to creating tides on Earth

It lights up the night sky

It's not always bright, but when it is the Moon can provide an excellent source useful not only for humans but for animals as well.

It keeps us safe from asteroids

The Moon plays a role in keeping us safe from impacts, either deflecting asteroids or taking hits, although the extent of this is debated.

It creates tides on Earth

Without the Moon, we wouldn't have lunar tides, one of the drivers for spreading nutrients through the ocean, and life would struggle to survive.

It keeps our planet stable

One of the Moon's most important facets is that it stops our planet wobbling too much on its axis, keeping us stable as we orbit the Sun.

It slows down our planet

Without the Moon, our planet would rotate a couple of seconds faster every 100,000 years. Over Earth's history, this would amount to a day shorter by several hours.

The Keck Observatory in Hawaii is involved in the hunt for exoplanets – and by extension their moons

that could indicate a dangerous volcanic moon with a molten surface of lava – almost a lunar version of the super-Earths that orbit their stars at extremely close proximities. The team behind the discovery, who originate from the University of Bern, suspect that the exomoon orbits a hot Jupiter that orbits its star with a period of just three days. Evidence of the exomoon was provided by observations of sodium gas around WASP-49b, located 635 light years from Earth, at an anomalously high altitude. The astronomers also discovered a further four systems that could host similar evaporating exomoons. As is the case with the other exomoon candidates, the confirmation of these exomoons may well rely on further investigations of the planetary systems in question.

The James Webb Space Telescope (JWST) will be instrumental in this. With projects like the JWST opening up opportunities to discover exomoons in distant planetary systems, one possible outcome of a deeper study of exomoons may be the revision of what we think of as a moon.

As exomoon discoveries continue, the definition of what we call a moon could become blurred in a similar way to how brown dwarfs have caused us to question the boundary between planets and stars. Often described as failed stars due to the fact they lack the mass to trigger the nuclear fission processes that define stars, brown dwarfs could possess bodies that orbit them. But if brown dwarfs are classed as planets rather than stars, shouldn't we then define the bodies in orbit around them as moons? That was a major question raised by a team of researchers who in 2019 collected data that suggested two tremendously massive exomoons are orbiting two distant brown dwarfs.

One of these brown dwarfs has around 10 or 11 times the mass of Jupiter and seems to be orbited by an exomoon around the same mass as the king of the Solar System, while the other brown dwarf has a mass around 13 to 14 times that of Jupiter with an exomoon that's almost five times the gas giant's mass. The planetary system discovered by the team, including University of Padova researcher Cecilia Lazzoni, who presented the results at the fourth Extreme Solar Systems conference in Reykjavik, Iceland, in 2019, is like nothing researchers have ever discovered before. If these potential exomoons can be confirmed, their immense mass means researchers will have to decide if they fit the definition of a moon at all.

Robert Lea
Space science writer
Rob is a science writer with a degree in physics and astronomy. He specialises in physics, astronomy, astrophysics and quantum physics.

What are planets like on the inside?

Even among the worlds of our Solar System we see a huge variety of planets

The terrestrial planets – Mercury, Venus, Earth and Mars – are separated into similar layers: a hot dense core, a warm pliable mantle and a cooled rocky crust. Mercury is around 70 per cent metallic and 30 per cent rocky. Its core is thought to comprise as much as 85 per cent of the planet, a liquid heart of iron 4,000 kilometres (2,485 miles) in diameter. This is covered by 600 kilometres (373 miles) of silicon-rich mantle and between 100 and 200 kilometres (62 and 124 miles) of rocky crust. Venus was expected to have a similar structure to Earth, but Venus has little magnetic field. Earth's field is created by rotation and convection in our molten core; Venus does seem to have a molten core, but for some reason it doesn't circulate in the same way. One possibility is that Venus' crust doesn't get recycled like Earth's does, so the whole core may be of uniform temperature as heat is not escaping to the surface.

Fortunately for us, Earth has a significant magnetic field that protects us from solar radiation, but our core isn't just liquid. The pressure at the centre of Earth is sufficient enough that the iron collected there becomes solid, despite the temperature being around 6,000 degrees Celsius (10,832 degrees Fahrenheit). This has been determined by studying the way seismic waves travel through Earth. The solid inner core is around 1,220 kilometres (760 miles) in diameter. This is surrounded by a liquid layer that is 2,200 kilometres (1,367 miles) deep, which is then topped by 2,900 kilometres (1,802 miles) of mantle and an average crust of 35 kilometres (22 miles).

Mars is also differentiated into layers, with a liquid core, a mantle – which appears to have driven volcanoes in the past – and a rocky crust. Like Venus, there must not be convection in the core, as Mars has no magnetic field.

Jupiter is the first of the gaseous planets and the largest in the Solar System. When it comes to gas giants, there's no sharp dividing line between the planet and its atmosphere. It's thought that Jupiter has a dense core, possibly rocky, surrounded by 'metallic' hydrogen. This is a strange condition predicted to occur under huge pressures, where hydrogen behaves like a dense electrically conductive substance. This layer is thought to cover 78 per cent of the thickness of the planet. It's thought that above this, normal liquid hydrogen smoothly fades into the gaseous hydrogen atmosphere. Saturn is thought to be similar to Jupiter, with a rocky or icy core surrounded by the same types of hydrogen layers.

Uranus and Neptune are called the ice giants, not because they have ice in the sense we know, but because a layer of mixed methane, water and ammonia surrounds their rocky cores – equivalent to the mantle in terrestrial planets. These are known as ices, though they form a hot, dense liquid near the core and fade out into the atmosphere without a defined surface.

Expert: Robin Hague

Robin is a science writer, focusing on space and physics. He is head of launch at Skyrora, coordinating launch opportunities for Skyrora's vehicles.

UNDER THE SURFACE

- ROCK
- MOLTEN ROCK
- ICE
- MOLTEN IRON
- IRON
- LIQUID METALLIC HYDROGEN
- HYDROGEN GAS
- ATMOSPHERE

MERCURY

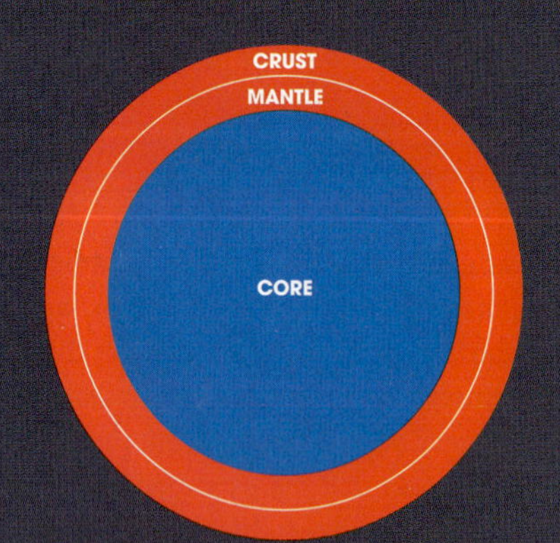

JUPITER

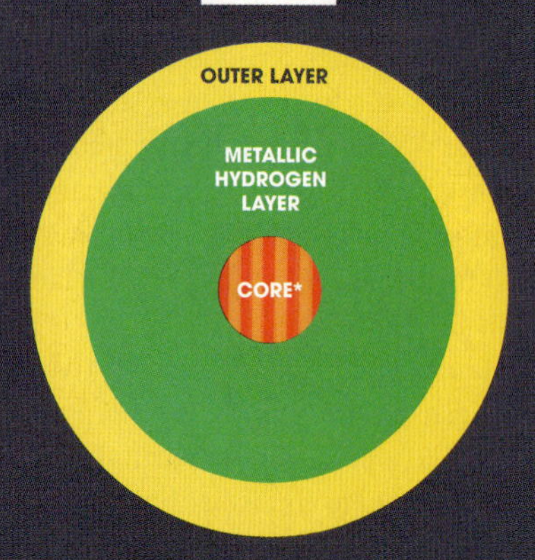

*JUPITER'S CORE REMAINS A MYSTERY TO SCIENTISTS, BUT IT'S HOPED THAT THE JUNO MISSION WILL SHED LIGHT ON ITS SIZE AND COMPOSITION

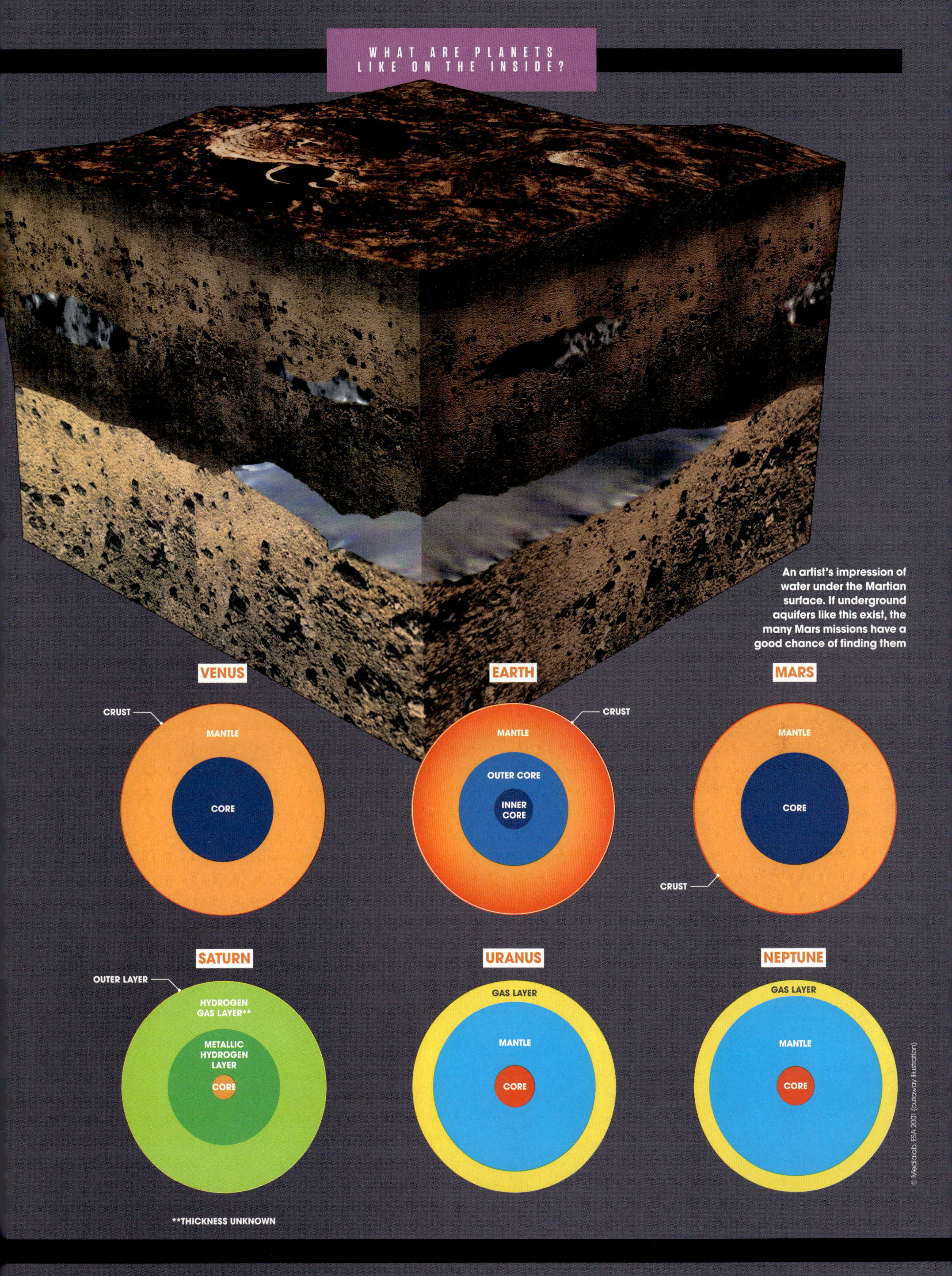

An artist's impression of water under the Martian surface. If underground aquifers like this exist, the many Mars missions have a good chance of finding them

Extreme stars

From the biggest and brightest to the smallest and dimmest, Andrew May takes a look at some stellar extremes

The Sun produces heat and light through nuclear fusion, converting hydrogen into helium. In this respect, it's a pretty typical star. Around 90 per cent of all stars are undergoing the same process, referred to as the main sequence of the stellar life cycle. Even so, there are some striking exceptions, such as stars of very low or very high mass or ones that have exhausted all their nuclear fuel. Here we take a look at some of these extreme stars – from brown dwarfs to supergiants and from neutron stars to weird hybrid stars. But first it's worth reviewing the basics of stellar evolution.

Although the stars in the night sky look similar to the naked eye, there's actually a wide variety of stellar types. This is partly because we see different stars at different points in their evolutionary cycles. This proceeds much too slowly for us to observe directly, so each star is like a single snapshot along the evolutionary path. It begins in a cloud of interstellar gas, where knots can form with sufficient mass that they start to collapse under their own gravity. As the collapse proceeds, the material gets hotter and denser, eventually forming protostars. Not all protostars are equal, even those formed in the same interstellar cloud at the same time. They come in a wide range of masses, from much smaller than our own Sun to many times larger. What happens next depends on the mass of the protostar. All but the very lowest mass stars soon become hot enough for nuclear fusion to take place, putting them on the main sequence. Somewhat paradoxically, however, it's the high-mass stars that burn through their nuclear fuel most rapidly before moving on to the later – and often much more dramatic – phases of stellar evolution.

Andrew May
Space science writer
Andrew holds a PhD in computational astrophysics and has written books on space and related subjects.

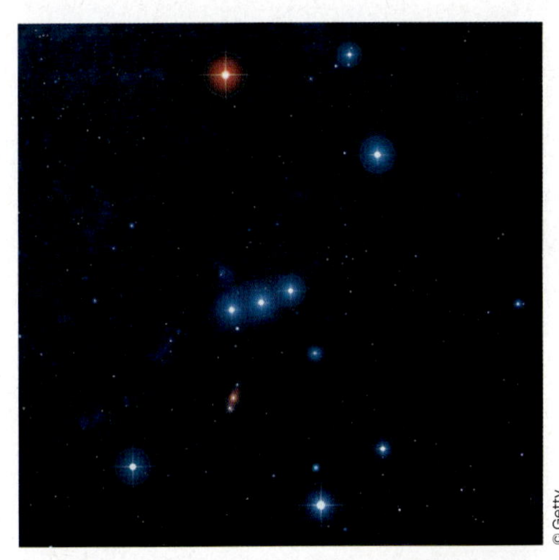

Two supergiants make up Orion: Betelgeuse at the top (red) and Rigel at the bottom (blue)

Brown dwarfs

Shining only very dimly, these low-mass objects are sometimes classed as failed stars

Brown dwarfs are the smallest stars, with masses in the range of 15 to 75 times the mass of the planet Jupiter. Like an ordinary star, a brown dwarf starts life by collapsing under its own gravity from a cloud of interstellar gas, but it doesn't have enough mass for the core temperature to rise to the point where hydrogen undergoes nuclear fusion. In other words, it never reaches the main sequence phase of stellar evolution. So why is a brown dwarf a star and not simply a very large planet? The reason is that while it's insufficient for ordinary hydrogen fusion, the core temperature is high enough for another kind of fusion involving a scarcer isotope called deuterium. This means the brown dwarf shines, albeit very dimly, with its own light – something a planet can't do. Even after all the deuterium is used up, the brown dwarf's retained heat means that it still radiates more energy than a planet. As it slowly cools down, this radiation declines from reddish light similar to a more conventional hydrogen-burning red dwarf star to very dim infrared light that is only barely perceptible, even with a powerful telescope.

Left: As this artist's conception shows, brown dwarfs may look more like planets than stars

Above: Comparative sizes of the Sun, a red dwarf, a brown dwarf, Jupiter and Earth

Brown dwarf
Star mass:
0.01 to 0.08 solar masses
Star diameter:
0.06 to 0.12 solar diameters
Star lifetime:
Trillions of years

Blue supergiants

These are extremely massive young stars that have already used up most of their hydrogen fuel and are now fusing helium into heavier elements. Despite their size and luminosity, they're actually very rare, accounting for less than one in a thousand of the stars in the galaxy. There are two reasons for this. To start with, when clusters of stars form in interstellar gas clouds, the distribution of masses is strongly biased towards low-mass stars. Secondly, high-mass stars have very short life spans – only about 10 million years compared to 10 billion or more for our Sun.

Blue supergiants are easiest to find in young star clusters like this one

Supergiant stars

The brightest stars in the galaxy live fast and die young

These stars are classified according to two main parameters: intrinsic luminosity and colour. In these terms, a supergiant is simply any star that lies at the upper end of the luminosity range. But there's actually quite a difference between blue and red supergiants. A blue supergiant is a comparatively young star that's so incredibly massive that it burns through all its hydrogen – and hence exits the main sequence – after just a few million years. On the other hand, a red supergiant is a star of somewhat lower mass, though still more than ten times that of our Sun, that's nearing the end of its life. After passing through the red giant phase of its evolution, it begins a final phase of carbon-forming nuclear fusion, causing it to expand to truly enormous dimensions.

Supergiant
Star mass:
Over ten solar masses
Star diameter:
30 to 500 solar diameters
Star lifetime:
10 million years

Red supergiants are gargantuan in size

Neutron stars

The densest stars of all are more massive than the Sun, but only miles across

Normal atoms are composed of electrons, protons and neutrons, but if they're squashed together forcefully enough then the protons and electrons combine to form further neutrons, and eventually that's all that's left. The result is one of the most extreme forms of matter known to science, and it's what neutron stars are made of. A neutron star is one of the possible outcomes when a star finally runs out of nuclear fuel and collapses under its own gravity. Small and medium-sized stars like the Sun end up as white dwarfs, while the most massive stars become black holes. In between the two, the end result is a neutron star – an object more massive than the Sun but compressed down into a volume the size of a city. Under normal circumstances it would be virtually impossible to observe a neutron star. But fortunately for astronomers, some of them give their presence away in a highly dramatic form as a pulsar. These are rapidly spinning neutron stars with strong magnetic fields which emit high-energy beams that can be detected as regular flashes of radiation when they periodically point towards Earth.

A NASA visualisation of a pulsar showing its magnetic field lines and emitted light beams

Neutron star

Star mass:
1.4 to 3.2 solar masses

Star diameter:
9.9 to 29.9 kilometres (6.2 to 18.6 miles)

Star lifetime:
Billions of years

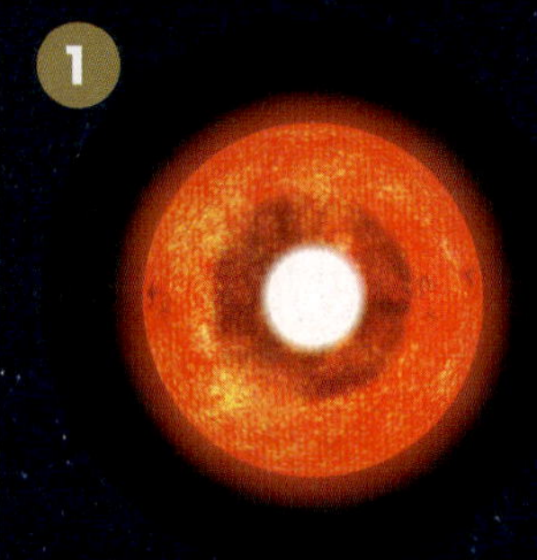

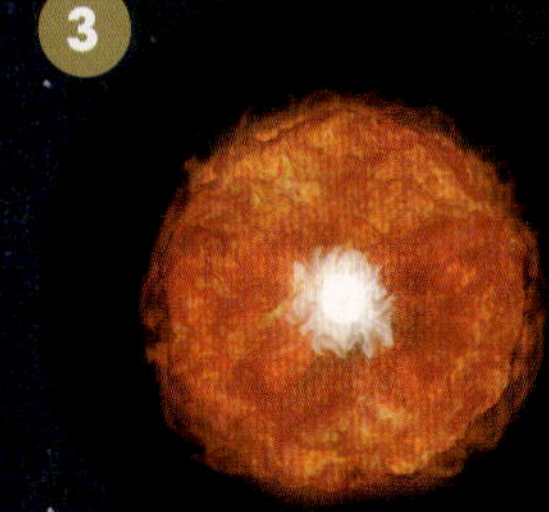

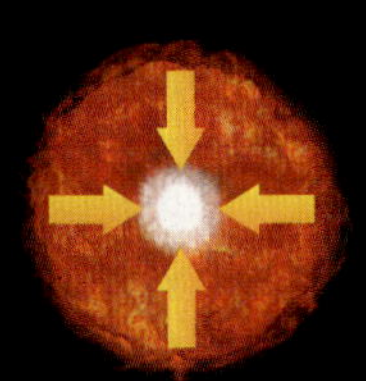

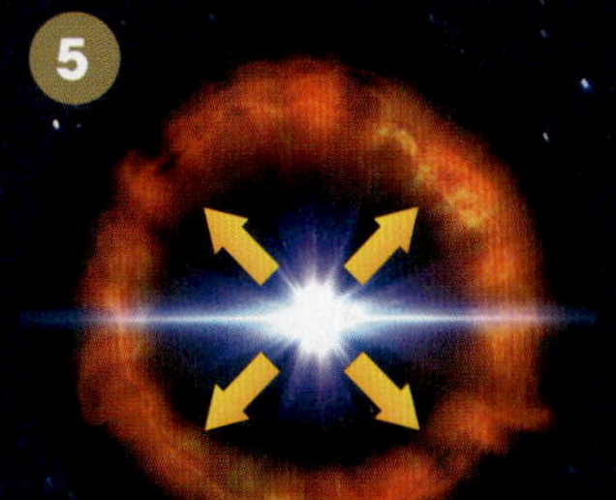

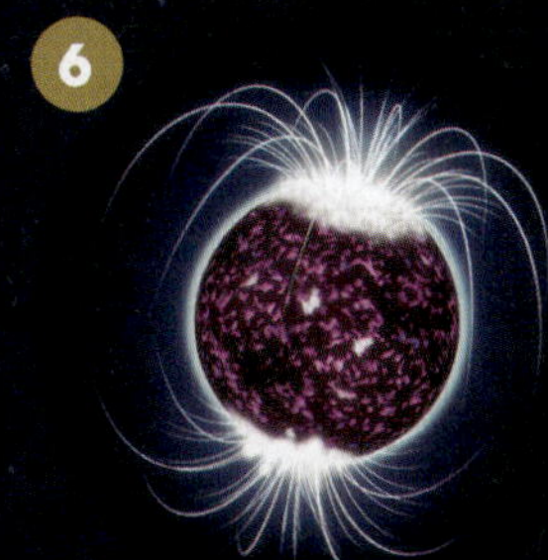

The birth of a neutron star

1 Massive star
For stars in this class, the final stages of main sequence evolution produce an iron-rich inner core.

2 Core collapse
When all the nuclear fuel has been used up, the dense core collapses under its own gravity.

3 Neutron formation
As gravity compresses the atoms in the core, electrons and protons are squashed together to form neutrons.

4 Outer layer collapse
The star's outer layers also start to collapse, falling into the core at a quarter of the speed of light.

5 Supernova explosion
The energy generated by the collapsing core blows off the outer layers in a spectacular supernova explosion.

6 Neutron star
All that remains is the incredibly dense core, composed almost entirely of neutrons.

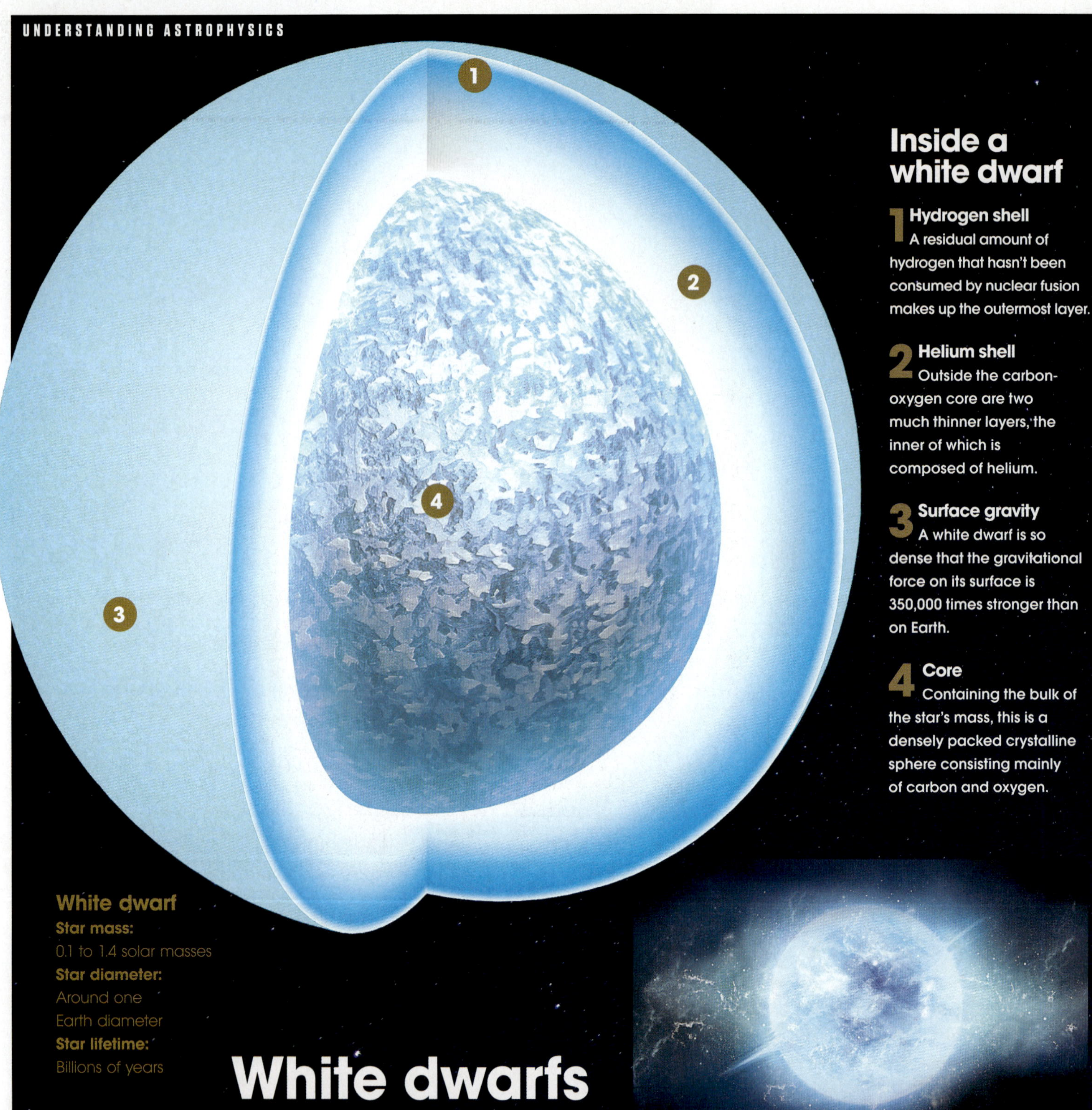

Artist's impression of a newly collapsed, superdense and very hot white dwarf star

Inside a white dwarf

1 Hydrogen shell
A residual amount of hydrogen that hasn't been consumed by nuclear fusion makes up the outermost layer.

2 Helium shell
Outside the carbon-oxygen core are two much thinner layers, the inner of which is composed of helium.

3 Surface gravity
A white dwarf is so dense that the gravitational force on its surface is 350,000 times stronger than on Earth.

4 Core
Containing the bulk of the star's mass, this is a densely packed crystalline sphere consisting mainly of carbon and oxygen.

White dwarf
Star mass:
0.1 to 1.4 solar masses
Star diameter:
Around one Earth diameter
Star lifetime:
Billions of years

White dwarfs

The surprisingly extreme fate of an ordinary star like the Sun

A white dwarf represents the final stage in the evolution of a star like the Sun, following on from the comparatively brief red giant phase. During the previous phase, the core of the star was gradually shrinking and getting hotter while the outer layers expanded to enormous proportions. Eventually the core generates so much energy that it blows the outer layers off completely, leaving just the dense, hot core behind. This core is the white dwarf, which will continue to exist, gradually losing its accumulated heat, for many billions of years. Although a white dwarf is extremely dense by ordinary standards, it's not as dense as a neutron star, where all the matter has been compressed down to neutrons. While a neutron star packs all of a star's mass into a volume just a few miles across, a white dwarf is around the same size as Earth. It's still composed of atomic nuclei and electrons like ordinary matter, but squashed down to the point where the separation between electrons is comparable to their wavelength. This produces a phenomenon called degeneracy pressure, supporting the white dwarf against any further collapse.

Red giant
Star mass:
0.3 to 10 solar masses
Star diameter:
20 to 100 solar diameters
Star lifetime:
Over 100 million years

© ESO/M. Kornmesser

Red giants

A dramatic phase of stellar evolution that lies in the Sun's own future

Most of the extreme stars we've listed here differ from our Sun in that they started out with lower masses, like brown dwarfs, or higher masses, like supergiants and neutron stars. The Sun's evolution will never take it in any of those directions. On the other hand, there are two extreme phases of the stellar life cycle that do lie in the Sun's own future: the red giant phase and the white dwarf phase. Currently the Sun is on the main sequence, converting hydrogen into helium via nuclear fusion. This phase is expected to end in about 5 billion years, after which the Sun will expand into a red giant so huge that it engulfs the planets Mercury and Venus – and possibly even Earth. The energy for this expansion, which is of the star's outer layers only, actually comes from a contraction as the core shrinks down and heats up.

A star within a star?

We speak to astrophysicist Cole Miller about hypothetical hybrid stars known as Thorne–Żytkow objects

Cole Miller is a professor of astronomy at the University of Maryland

What is a Thorne-Żytkow object?

A hypothetical object in which a red giant or supergiant with a radius on the order of the Earth-Sun distance contains a neutron star in its core. Neutron stars and regular stars can orbit each other, and the suggestion is that when the regular star evolves into the red giant phase it might sometimes swallow the neutron star.

What's the history behind this strange stellar idea?

An early suggestion was by the great Soviet physicist Lev Landau, who proposed in 1938 that a small neutron star core at the centre of normal stars could provide their power sources. That turns out not to work, but in 1977 Kip Thorne and Anna Żytkow of the California Institute of Technology made the suggestion that bears their names. Landau hoped that the idea would be amazing enough to save him from arrest as a dissident in Stalinist Russia, but it didn't work and he spent a year in prison.

Are these objects purely theoretical, or have they actually been observed?

The evidence isn't clear because differences from standard red giant stars are subtle. As Thorne and Żytkow noted, such stars "are thoroughly hidden from the prying eyes of the astronomer by the huge, tenuous red-giant envelope". It could be that the abundances of certain isotopes would be different than normal in such stars. There was a report in 2014 that a star named HV 2112 has anomalous abundances of the element rubidium and the expected very high luminosity – about 100,000 times that of the Sun. However, in 2018 a reanalysis by another group found a lower luminosity and no excess of rubidium. At the same time, the 2018 paper proposed its own stellar candidate, HV 11417, so there is still hope.

© Lisa Helfert, Cole Miller (headshot); Getty (white dwarf cutaway)

Dark matter: where did it come from?

It's eluded us for decades, but bubbles could be the answer to the universe's most mysterious substance

If something cast a shadow but you couldn't see it, you'd be intrigued. If it did that on the scale of the universe, you'd be perplexed. If you then couldn't find it despite your best efforts, you'd likely be frustrated. This is the situation that scientists looking for dark matter find themselves in. But dark matter doesn't even cast shadows; it's invisible to all electromagnetic radiation. And yet we have solid evidence that it exists. In fact, current calculations show that it comprises 80 per cent of all of the mass in the universe. But a 2020 study modelling dark matter as filtering through bubbles after the Big Bang may help point scientists in a new direction.

Although the mystery of dark matter can be traced as far back as Lord Kelvin, the modern breakthrough in research came in 1933. Fritz Zwicky was a Swiss astronomer working at the California Institute of Technology. Averaging the rotations of galaxies in the Coma Cluster, a large cluster of over a thousand galaxies, Zwicky noticed that their speeds were excessive. So much so, in fact, that the galaxies should have flown apart. There wasn't enough visible matter to gravitationally bind each galaxy together. He concluded that there must be a 'dunkle materie' – or 'dark matter' – component to them, which he calculated to be 400 times what could be seen. Modern calculations have lowered this value.

Six years later, American doctoral student Horace Babcock showed dark matter present in the Andromeda Galaxy. Curiously, it was more concentrated in the periphery of the disc than the centre. This was the first indication that galaxies had dark matter 'halos'. Then, in the late 1970s, American astronomers Vera Rubin and Kent Ford showed that galaxies contained between five and ten times more dark matter than luminous matter.

"The puzzle of dark matter forces us to think in new ways"
Michael Baker

Fritz Zwicky was the first scientist to measure the extent of dark matter in other galaxies

Where did dark matter come from?

Bubbles in the early universe may have solved the mystery

1 Big Bang occurs
The universe begins as an extremely hot, dense state.

2 Inflation happens
At 10-36 seconds the universe expands with unimaginable speed in a period known as inflation. This cools the universe further.

3 Thermal freeze out
Very quickly, the subatomic particles of ordinary matter 'freeze' out of the dense plasma as it cools. So does dark matter.

4 Bubbles form a particle
called phi starts a phase change in the universe. These act as nucleation sites, growing as expanding bubbles.

5 Filtering of matter
Despite particles freezing out, they're still hot, including dark matter. Some of the latter pass into the bubbles and gain mass.

6 Bubbles merge
The bubbles merge, and the phase change completes. All the surviving dark matter particles are now part of the universe.

Gravitational lensing in the Abell 1689 cluster, caused by its dark matter

The term 'dark matter' is a slight misnomer. It isn't simply dark; it's completely invisible, no matter which part of the electromagnetic spectrum it's observed in. Its only observable effect is gravitational, much like how wind is only visible when it affects leaves, branches and hair. But this gives dark matter a dramatic quality, one predicted by Einstein and one that serves as another piece of evidence of its existence.

In 1915 Einstein presented his general theory of relativity, which models space and time – or just 'space-time' – in our universe as behaving like a stretched fabric warped by the presence of mass or energy. This warping of space-time in a mass' vicinity is the explanation for gravity. This is true of planets, stars, galaxies and entire galaxy clusters.

Multiple experiments have confirmed the theory with incredible accuracy. Although not the first to do so, in 1937 Zwicky realised the implications of Einstein's theory. The path of light would be curved by this warped space-time, meaning that whole galaxy clusters could be used as gravitational 'lenses'. It was another 42 years before this effect was observationally

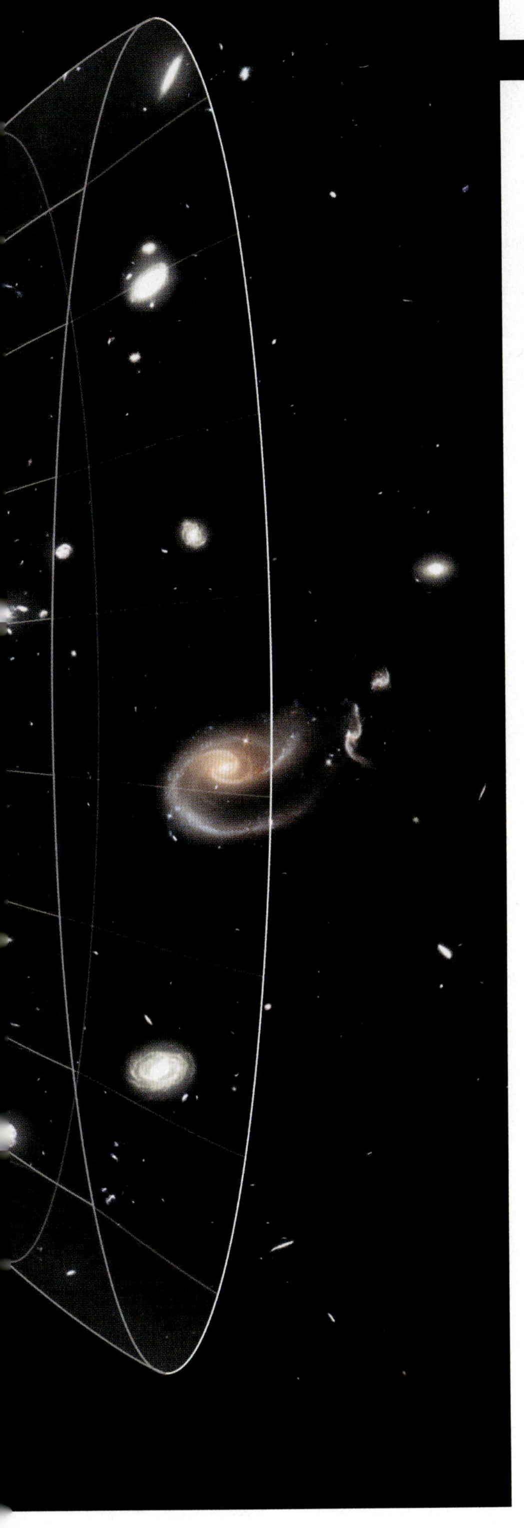

The quasar and its double image QSO 0957+561, detected in 1979, proved gravitational lensing

confirmed with the twin quasar QSO 0957+561 A/B, gravitationally lensed by the galaxy YGKOW G1. Since then gravitational lensing has become a tool to measure the extent of dark matter, such as in the galaxy cluster Abell 1689. But where did it come from? To answer that, we have to go back to the very beginning.

The universe started 13.8 billion years ago in an extremely hot, dense state, known colloquially as the Big Bang. This intense region, containing all of the energy of our entire universe, was far too hot and dense for matter to exist in any meaningful sense. There was constant creation and annihilation of particles. Then things happened very rapidly.

Between 10^{-36} and 10^{-32} seconds after the Big Bang, in a period known as inflation, the universe expanded 10^{26} times. Quarks and subatomic particles formed in the subsequent fractions of a second that followed. However, it was still too hot for protons and neutrons, which comprise ordinary atomic nuclei, to form. According to a lot of scientists, as expansion continued and the temperature dropped, subatomic and dark matter particles both condensed out of the cosmos' plasma, like ice crystals forming in a cloud.

These particles are often called 'thermal relics', since they are the leftovers of a higher temperature universe. The main example of thermal relic dark matter candidates are Weakly Interacting Massive Particles (WIMPS), which have been the source of most searches for this elusive substance.

But according to a new study, an alternative scenario may have played out. Instead of freezing out of cooling plasma, dark matter may have filtered out into bubbles that formed during what's known as a phase transition, when the universe changes its state from one form to another as it continues to expand. Examples of phase transitions abound in physics, including the early universe. One

What is dark matter?

It's the majority of matter

Invisible dark matter makes up 80 per cent of the mass in the entire universe.

It affects light

Because of general relativity, dark matter acts as a gravitational lens and 'bends' light.

It could be a new particle

Dark matter could be an undiscovered particle that hardly interacts with normal baryonic matter or radiation.

It could be a WIMP

Weakly Interacting Massive Particles, formed after the Big Bang, could be one explanation.

It could be an axion

Evidence for WIMPs is weakening, giving ground to axions as a possible alternative.

Newton's laws may need modifying

One alternative to dark matter is tweaking Newton's laws so that they behave differently at galactic edges.

It forms the universe's structure

Scientists think galactic clusters form in an existing 'dark matter web', forming the universe's large-scale structure.

It's long lived

Axion dark matter could last 3,000 trillion years – roughly 220,000 times the age of the universe.

Dark matter vs Dark energy

Even more mysterious than dark matter is a phenomenon in the universe called dark energy

72% Dark energy

23% Dark matter

5%

Everything else, including all stars, planets and us

Invisible matter

Scientists think that dark matter may be like ordinary matter, only invisible and non-interacting. However, we don't know. We could be completely mistaken, and it could be something else entirely, yet to be theorised.

It has a gravitational effect

Like ordinary matter, its dark counterpart has a gravitationally attractive effect on its surroundings. This can be seen in the gravitational lensing of large galaxies and galaxy clusters.

It was seen in the 19th century

Lord Kelvin first noted the peculiar motions of stars in the Galactic Centre, concluding in 1884 that a large number of 'unseen dark bodies' were present and causing the gravitational effects.

Voyager 1 has disproved one theory

The Voyager 1 probe, now outside the Sun's heliosphere, would have been able to detect faint radiation from super-small black holes, a possible dark matter candidate. None was detected, disproving the theory.

It's visible in the Big Bang's echo

The cosmic microwave background, the relic radiation of the early universe, can show patterns relating to matter distribution. In particular, it can show how the universe's web of dark matter evolved over time.

It's speeding things up

In 1998, results from two separate instruments – one of which was the Hubble Space Telescope – showed that 5 billion years ago, the expansion of the universe started accelerating. The cause is labelled 'dark energy'.

It's most of the universe

According to data from NASA's Wilkinson Microwave Anisotropy Probe spacecraft, dark energy comprises an astonishing 72 per cent of the energy density of the entire universe. Dark matter comprises 23 per cent.

Einstein's mistake may not be wrong

Einstein's equations showed the universe was expanding when it was thought to be static. He introduced the cosmological constant to fix this. Later admitting it was a mistake, it may nonetheless explain dark energy.

Then again, it may be

From calculations, the cosmological constant is 120 orders of magnitude too large to explain observations. Fortunately there are other models for dark energy, including a proposed new fundamental force called quintessence. But there's no evidence yet.

We're trying to solve the mystery

The Dark Energy Survey (DES) is just one of many projects trying to uncover dark energy's secrets. In DES' case this involves mapping millions of galaxies and thousands of supernovae to see what they reveal.

The IceCube Observatory hunts for neutrinos below the ice in Antarctica

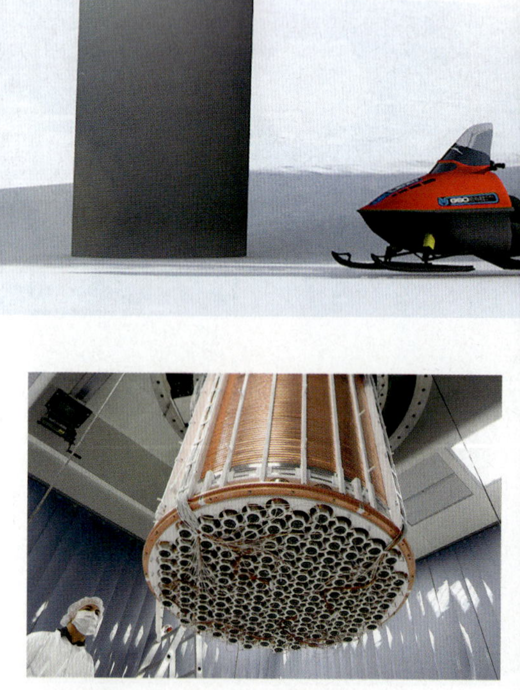

The XENON1T experiment detected an excess signal, which currently remains unexplained

everyday example is boiling water. As water boils it changes phase from liquid to gas. This change can be seen in the water itself as bubbles of vapour form before escaping. Something similar occurred fractions of a billionth of a second after the Big Bang.

In the new study, published in Physical Review Letters, a trio of physicists working in Australia, Switzerland, Germany and Texas propose an additional new particle called phi. "The phase transition occurs when the phi field has a non-zero value in the lowest energy

configuration – like the Higgs field today," says Dr Michael Baker of the University of Melbourne. In the early universe, these phi particles acted as nucleation sites, with their associated fields expanding like bubbles. Dark matter particles that formed in the super-hot, dense 'soup' of the universe collided with the bubble walls. Many simply bounced off, but some had enough momentum to pierce through into the bubbles. As they did so they gained more mass, because the particles were able to couple more strongly to the Higgs field, responsible for the Higgs boson.

The Higgs boson was discovered by the Large Hadron Collider (LHC) in 2012 and is what gives other particles mass. Due to conservation of energy, once inside these regions, now-massive particles slowed down, dramatically reducing particle-antiparticle annihilations and thus cementing their place in the universe. The bubble expansions continued apace until they merged, enveloping the entire universe and leaving behind only those dark matter particles lucky enough to get in.

No one has yet been successful in directly detecting dark matter, so how would someone look for dark matter filtered by cosmic bubbles? "We consider the best prospects for finding evidence for our proposal are collider searches for the new [phi] particle, detection of dark matter annihilation products or detection of background gravitational waves," explains Baker. In the 2020 study, the authors state that particle-antiparticle annihilations in the galactic disc could give rise to neutrinos on the order of 1015 electronvolts. Such an event was observed by the IceCube Neutrino Observatory in Antarctica. But these events are rare, and may also come from other sources, such as supernovae.

"Negative results are the mainstay of dark matter physics, but they rule out lines of enquiry"

Part of Washington University's ADMX haloscope, designed to detect dark matter axions

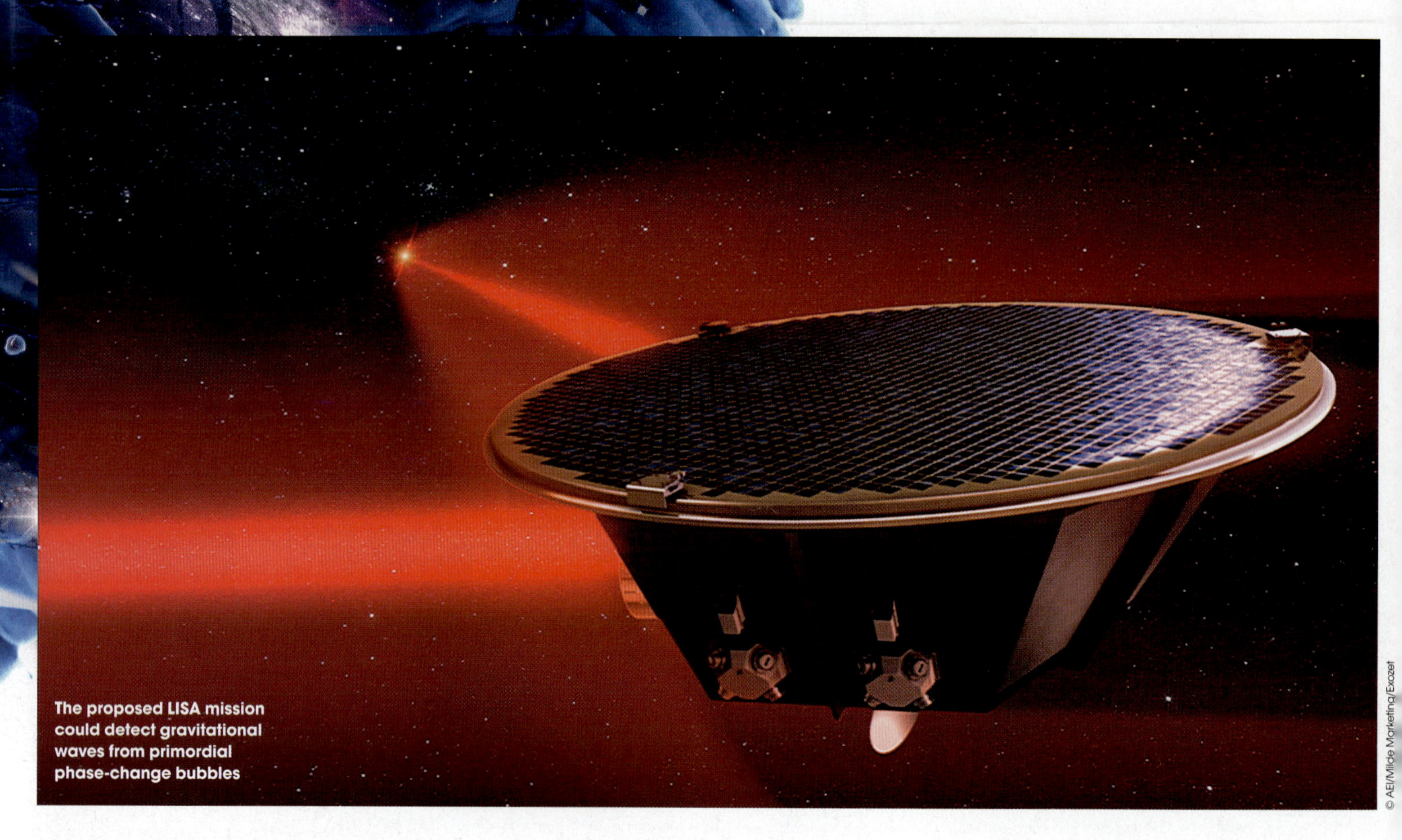
The proposed LISA mission could detect gravitational waves from primordial phase-change bubbles

What of colliders? In October 2020 the ATLAS collaboration at the LHC presented results using the Higgs boson itself as a tool for searching for dark matter. Rather than looking for dark matter directly, the ATLAS detector looked for signs of missing momentum that could indicate its presence. Nothing significant was detected, but the branching fraction – the ratio of particles decaying into another particular one – was whittled down to a maximum of 0.11 – or 11 per cent – with 95 per cent confidence.

As Baker points out: "We used an earlier bound of 0.19 on the Higgs branching ratio, so that constraint is now a bit stronger, although our mechanism is still perfectly viable. Note that in our case it puts a bound on the new scalar particle [phi] in our model, which is not dark matter. The fermion 'chi' is, but the scalar would behave very much like dark matter if produced in a collider." So a detection of either would be a breakthrough. Such negative results are the mainstay of dark matter physics, but they are important for ruling out lines of enquiry and chasing up more promising leads.

What about the detection of gravitational waves – ripples in the fabric of space-time – that Baker mentions? The study specifically mentions possible use of the Laser Interferometer Space Antenna (LISA), a space-borne gravitational-wave observatory using laser interferometers spaced 2.5 million kilometres (1.5 million miles) apart. Gravitational waves were first detected by the ground-based LIGO (Laser Interferometer Gravitational-Wave Observatory) and Virgo collaboration in 2015. Due to launch in 2034, early universe phase transitions are one of LISA's science goals. Due to the expansion of the universe over time, any gravitational waves resulting from bubble formation would now have a huge amplitude, necessitating LISA's large span.

And what of the dark matter itself? Could the chi particles be WIMPS, or something else? As they 'froze out' from the cosmic plasma, WIMPS would be considered thermal relics. "Most of the major current experiments are searching for dark matter based on the hypothesis that it was produced in this way," says Baker. The bubble-filtered dark matter of the study didn't form this way, however. "Recently attention has begun to shift focus to other production mechanisms; axion dark matter is an example of a non-thermal relic that is currently actively being searched for," he says.

Axions are a hypothetical particle first proposed to explain an unusual problem in quantum physics, and are now also a possible dark matter candidate. One operation looking for them is the Axion Dark Matter Experiment (ADMX) at Washington University. Some axions are supposed to turn into microwave photons

The Coma Cluster's galactic rotation curves show the presence of large amounts of dark matter

in strong magnetic fields, so ADMX consists of a microwave cavity inside a superconducting magnet, called a haloscope.

So far the international team has ruled out low-mass axions. As Professor Gray Rybka of the University of Washington says: "ADMX is currently searching masses above 3.53 microelectronvolts, and we'll let the world know if we find anything! One interesting fact is that the expected axion signal, if we find it, should be very clear." But another axion detector has already seen something, which at the time of writing remains unexplained. The XENON1T detector, operated by INFN Laboratori Nazionali del Gran Sasso, Italy, recorded an excess signal between 2016 and 2018, first published in 2020. It's possible that it's due to the presence of tritium atoms in the ultra-pure xenon used, but nothing has been ruled in or out yet.

Follow-up experiments currently underway may provide a definitive answer. Something echoed by Baker, who's looking forward to following up his own work. "The experimental landscape is quickly changing, with huge progress having been made in the last 10 to 20 years. The puzzle of dark matter forces us to think in new ways, and even if these ideas turn out to be unrelated, they could provide the key to understanding some of the deepest mysteries in the universe."

Kulvinder Singh Chadha
Space science writer

Kulvinder is a freelance science writer, outreach worker and former assistant editor of *Astronomy Now*. He holds a degree in astrophysics.

The biggest unanswered questions about dark matter

Is there more than one kind?

Could there be an entire family of dark matter particles as opposed to just a single type? The Standard Model of particle physics says that ordinary matter is composed of electrons as well as the quarks that make up protons and neutrons. It also includes the fundamental force-carrying particles, such as photons, gluons, bosons and some other short-lived matter particles. Could there be a similar but separate model for dark matter?

Are there corresponding 'dark forces'?

Forget horror and fantasy here. If there are a range of dark matter particles, could they experience their own dark matter forces? If so, could that also sometimes affect ordinary particles? This is the hope of a collaboration working with the Super Proton Synchrotron at CERN. The team has examined billions of high-energy electrons, looking for an energy deficit in the resulting photons. The idea is that 'dark photons' would interact with normal photons and carry away some energy, thus showing themselves by their absence. Nothing's yet been detected, but the team will try again in 2021 after an upgrade.

Could it be responsible for the universe's matter?

In the very early universe, fractions of a second after the Big Bang, equal amounts of matter and antimatter should have been created. They then both should have annihilated one another, leaving a universe filled only with radiation. But the universe is filled with matter, with little sign of the missing antimatter. What led to this asymmetry is still one of the biggest mysteries in cosmology. Could dark matter be the culprit? The Baryon Antibaryon Symmetry Experiment at CERN is testing whether dark matter particles could interact preferentially with antimatter. This might account for the lopsidedness we see today.

Could it already have been detected?

The Gran Sasso National Laboratory in Italy hosts the XENON dark matter experiment, of which the XENON1T detector team announced an unexplained signal excess in June 2020. The possibility is that it could be due to contamination of the xenon by tritium, but it's not yet been confirmed. Gran Sasso also hosts the DAMA/LIBRA experiment, which looks for dark matter in the galactic halo using a different design based on scintillation crystals. This too detected unexplained signals, with seasonal variations. The results are inconclusive, but a new version called SABRE, based at Gran Sasso and Australia, may prove or disprove any known natural variations.

How did Earth get its water?

Moon rocks suggest that the water might have been here all along

Earth's water may have been here since the planet formed, and not delivered later by collisions with icy comets. New research analysed Moon rocks brought to Earth by the Apollo program and sheds light on our planet's earliest days. Although more than 70 per cent of Earth's surface is covered in water, overall our planet is relatively poor in water and other volatile molecules compared with most other bodies in the Solar System, said Lars Borg, a planetary scientist at Lawrence Livermore National Laboratory in California.

Scientists have long debated how Earth came to possess water. Two major scenarios prevail, and both involve ancient cosmic impacts, the most notorious of which saw the proto-Earth collide with a Mars-size rock dubbed Theia that helped give birth to the Moon. In the first scenario, "Earth is born dry and inherits its water through the addition of material from water-rich bodies such as comets and primitive meteorites," Borg said. "In the second, Earth forms with volatile element abundances that are similar to the Solar System average, but loses most of them during the giant impact that formed the Moon."

To help solve this mystery, scientists analysed rocks from the Moon collected during the Apollo missions. The researchers focused on levels of rubidium-87, which is volatile and radioactive, and strontium-87, the stable product of its radioactive decay. They found that levels of strontium-87 were relatively low in lunar highland rocks that crystallised around 4.35 billion years ago, suggesting that levels of rubidium-87 and other volatiles were similarly low in Earth and the Moon when they formed. But that conclusion contradicts both of the major theories for how Earth got its water.

"Our work suggests that Earth and the Moon formed with about the same amounts of volatile elements as they have today," Borg said. "This doesn't mean that no water was added to Earth by comets and meteorites, but simply that the majority of water was inherited from the materials that Earth and the Moon formed from." These findings suggest that both Theia and the proto-Earth were strongly depleted in volatile compounds before the massive collision. This hints that both bodies formed in the inner Solar System relatively late in the Solar System's history, after 4.45 billion years ago, when the young Sun's heat would have baked many of the volatiles out of these bodies.

These findings may also help explain other mysteries regarding the origin of Earth and the Moon. "These include explaining the baffling observation that Earth and the Moon have similar oxygen, chromium and titanium isotopic compositions when most formation models predict they should be different," Borg said.

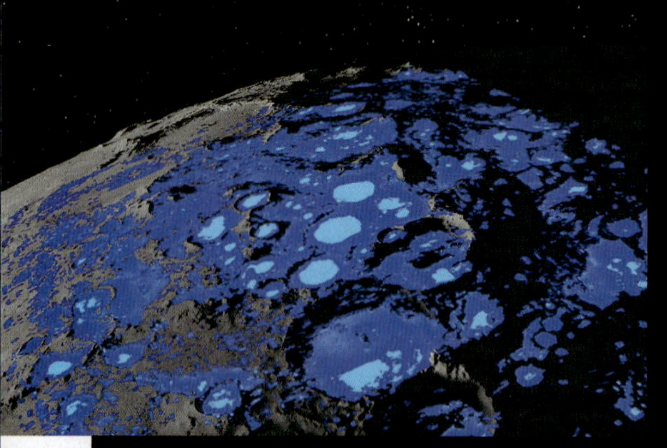

A map of possible water at the lunar south pole based on data from a NASA lunar orbiter

Where is water on the Moon?

A Watery chemistry in Clavius crater

The detection of water molecules in Clavius, one of the largest craters visible from Earth, hinted at some degree of surface-wide hydration of the lunar regolith.

B Dirty polar ice

The Lunar and Planetary Institute's Paul Spudis used Chandrayaan-1's radar data to estimate that 600 million tonnes of water ice could be locked up in just 40 polar locales.

C Mapping the poles

The pole-to-pole orbit of Chandrayaan-1 allowed NASA's onboard Moon Mineralogy Mapper instrument to map the surface ice across the Moon's southern pole.

D A wispy, watery atmosphere

It's believed that water molecules freed from lunar regolith grains by the warming Sun might bounce across the surface, travelling through the Moon's thin exosphere.

E Volcanic water

In 2017, Hawaii University's Shuai Li found evidence of water in large pyroclastic deposits, including those found near the Apollo 15 and 17 landing sites.

How the Moon was made

1 Theia approaches the proto-Earth
A Mars-sized object entered an unalterable collision course with the early Earth, forming the Moon that we see today.

2 Earth takes a big hit
The impactor hit Earth in a head-on collision, vaporising both Theia and Earth's mantle as it struck.

3 Material gets thrown out
The vaporised material from both bodies mixed and was thrown outwards by the huge impact.

4 Debris begins to gather
Smaller objects began to condense out of the vapour while continuing to orbit around Earth.

5 The Moon takes shape
Many of the smaller objects stuck together to form a proto-Moon in orbit around Earth.

6 Our companion is formed
Eventually all the pieces came together to form the basis of the Moon we see today.

The proto-Earth had many calamitous encounters with other bodies

> "Earth and the Moon formed with about the same amounts of volatile elements as they have today"
> **Lars Borg**

Celebrating Hubble

The Hubble Space Telescope has astounded for decades, returning breathtaking images and groundbreaking science from its position high above Earth

"Hubble has completely transformed our view of the universe, revealing the true beauty and richness of the cosmos."

John Grunsfeld, former NASA astronaut

No mission has ever captured the public's imagination quite like Hubble. Since its launch on 24 April 1990, the venerable space telescope has made over 1.5 million observations of planets, star clusters, nebulae and galaxies that not only look stunning, but have furthered our understanding of the universe. From its position in orbit 535 kilometres (332 miles) above our planet, Hubble is free from the atmospheric distortion that plagues ground-based observatories, giving it a clear picture. But this wasn't always the case. The first images Hubble returned were distorted; astronomers concluded that the telescope's main mirror had been sent into space flawed. Corrective optics were devised, installed by a Space Shuttle mission in 1993, and from then on its place among the stars was cemented. Even with its successor, the James Webb Space Telescope, now operational, Hubble still has plenty to reveal to us.

Westerlund 2

23 April 2015

Stars burst to life in this mesmerising image, with topaz and amethyst-hued clouds of gas and dust lit up by the harsh radiation of the thousands of young stars speckled across the background sky. For Hubble's 25th anniversary, astronomers focused the telescope on a region known as Gum 29, 20,000 light years away in the constellation of Carina. The stellar firework just right of centre is Westerlund 2, a cluster of stars estimated to be just 2 million years old.

Messier 51 and NGC 5195

25 April 2005

Known as the Whirlpool Galaxy due to its glittering arms, Messier 51 is a grand-design spiral galaxy 25,000 light years away. Its sweeping swirls are where new stars are being born from dense hydrogen clouds, while its yellow centre is much older.

Jupiter and Europa

25 August 2020

Snapped while Jupiter was 653 kilometres (406 miles) away, this image highlights the tumultuous bands of storms that circle the planet in opposite directions. The bright-white storm in the upper left was brewing when the image was taken, allowing scientists to see a storm in its early stages. Also on display is the smallest of the planet's Galilean moons, Europa, an icy moon suspected to have a subsurface ocean.

V838 Monocerotis

3 February 2005

This red supergiant drew attention in 2002 when it mysteriously brightened for three months. The smoky swirls are a light echo – gas and dust reflected the light from the star's eruption, delaying it on its way to reach Hubble's instruments.

Carina Nebula

24 April 2007

This panoramic vista showcases 50 light years of the Carina Nebula in one majestic mosaic, moulded from 48 separate frames taken by Hubble. The billowing clouds in the region have been shaped by fierce stellar winds and radiation from massive young stars. These stellar giants grow huge and hot, but die young. Eta Carinae, a star on the left, has begun expelling material and is expected to go supernova relatively soon in astronomical terms, despite being only around 3 million years old.

"It's my favourite because of its beauty and because it contains in a single mosaic the full range of stellar lives, from their birth in dense clouds of dust and gas to the later stages where the dying stars expel matter at high velocities out into the interstellar medium"

Dr Kenneth Carpenter, Hubble operations project scientist

NGC 3603

2 October 2007

This nebula of ionised hydrogen contains one of the most populated young star clusters in the Milky Way, showcasing stars at different stages of their lives. Within the last million years, a collapse within the cloud caused thousands of stars to spring to life, all different shapes, masses and sizes – great for researchers looking into stellar evolution. The absence of nebulosity in the central region also allows for unobstructed views of the star cluster, making it a prime target for study. This window within the clouds was carved by radiation from the hot blue stars in the middle of the cluster.

Helix Nebula

16 December 2004

One of the brightest planetary nebulae looks like an eye looming in the darkness of space. These objects form when a star too small to go supernova sheds its outer layers as it runs out of fuel at the end of its life. As these gaseous layers spread out into the surrounding space, the star at the centre cools to become a white dwarf. These stellar remnants exude radiation, heating the expelled material. The Helix Nebula was the first planetary nebula observed to have cometary knots – dense filaments of gas that resemble comets with tails – dispersed within its rings of material.

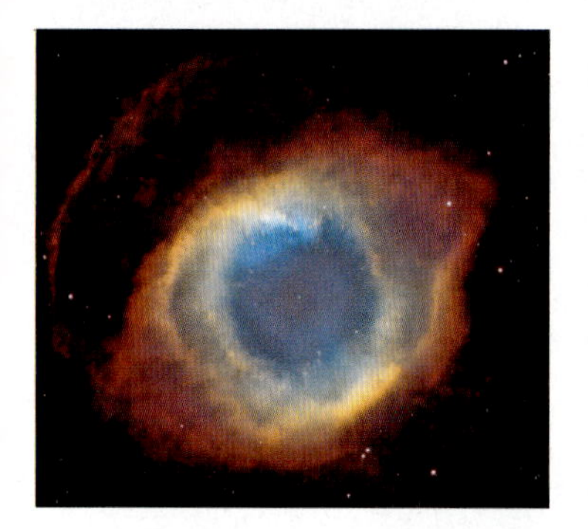

Saturn spokes

9 February 2023

With Saturn's orbit of the Sun taking over 29 years, the Ringed Planet experiences seasons much slower than we do here on Earth. The northern hemisphere is approaching its autumn equinox, which will take place on 6 May 2025, taking the planet into its so-called 'spoke season'. Two dark smudges can be spotted in the B ring to the left of the planet in this 2023 Hubble image. As autumn approaches, these spokes will become more plentiful and pronounced. These strange radial markings can appear light or dark depending on the angle they're lit from.

Hubble Ultra Deep Field

21 September 2006

Pointing Hubble in the direction of an 'empty' region of space, 800 exposures were collected over a period of 114 days to obtain this astonishingly detailed canvas of nigh-on 10,000 galaxies, with spirals, ellipticals and irregulars all accounted for. Those in red are some of the oldest galaxies in the universe, forming around 800 million years after the Big Bang – these only become visible to us when observing in near-infrared. The more typical spiral and elliptical galaxies are closer to us, and appear much brighter.

N159

9 September 2016

Like cosmic storm clouds littered with sparkling gems, glowing gas and dark dust lanes are juxtaposed against a backdrop of blue in this region of intense star formation in the Large Magellanic Cloud, a satellite galaxy of the Milky Way. This stellar nursery is the birthplace of massive stars, which irradiate the surrounding gas. The outflow from these young stars forms hypnotising patterns in the gas and causes these clouds to emit an eerie glow, lit up by stellar radiation. This region stretches across 150 light years and lies just 180,000 light years from Earth.

Mystic Mountain

23 April 2010

To celebrate Hubble's 20th year in orbit, the space telescope was turned towards these towering pillars of dust and gas resembling a magnificent mountain ignited with colour. Located 7,500 light years away in the Carina Nebula, each colour corresponds to the glow of a different element: oxygen is blue, sulphur is red and hydrogen and nitrogen are green. This is a very turbulent region of space, with stellar radiation constantly eating away at and moulding the mountain. Within the structure, new stars are being born from the compressed material. These juvenile stars fire off jets of gas, which are responsible for creating the wispy streamers observed at the peaks.

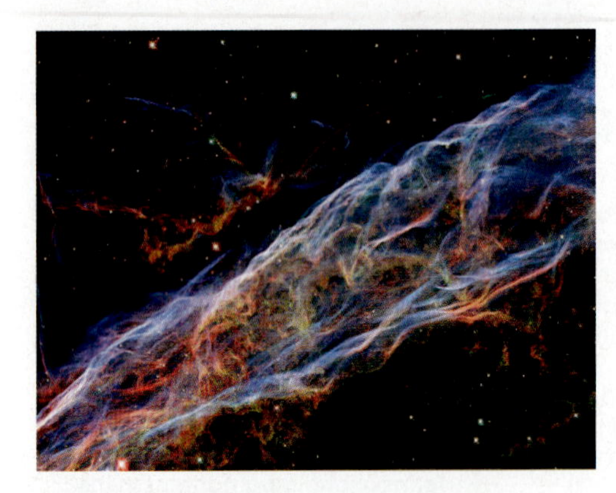

The Veil Nebula

29 March 2021

10,000 years ago, a star 20 times the mass of our Sun exhausted all of its fuel and gave out in a glorious supernova explosion. This violent death sent waves of its stellar material out into the cosmos, which over time settled into the delicate filaments that make up the Veil Nebula, part of a 2,100-light-year-distant supernova remnant called the Cygnus Loop. Five different filters were used to bring out the colours of the twisting ribbons of ionised gas. This only shows a small section of the nebula; the entire loop, also catalogued as NGC 6960, measures 110,000 light years across.

AG Carinae

23 April 2021

Though AG Carinae is one of the most luminous stars in the Milky Way, its distance of 20,000 light years means we can't really see it with the naked eye. The star is classed as a luminous blue variable due to its erratic behaviour, dimming and brightening unpredictably. Due to its unstable nature, the star is prone to eruptions. An outburst event around 10,000 years ago is believed to be the source of the ring of nebulosity surrounding the central star, suggesting that it may be struggling to hang onto its material as it approaches the end of its life.

NGC 7172

28 March 2022

With galaxies galore in any direction you can point a space telescope, astronomers are able to view these cosmic objects from every angle to form a full picture of their structures, classifications and properties. NGC 7172 is just one of many galaxies we get a side profile of, with the spiral galaxy's edge facing towards us. The dark dust lanes obscured any secrets hidden at the galaxy's centre, but spectral analysis revealed that NGC 7172 is a Seyfert galaxy, a special class of spiral with a supermassive black hole that's currently consuming matter from its accretion disc.

NGC 1300

23 April 2015

NGC 1300 faces almost directly towards us, showing off the galaxy's central bar in exquisite detail. Rather than curling all the way to the central nucleus, a barred spiral galaxy's star-saturated arms connect to the ends of a bright bar that runs through the galactic centre. It's estimated that 60 per cent of galaxies are spirals, and around two-thirds exhibit a central bar, making them the most commonplace – in fact, our galaxy is classified as one. Galaxies with larger bars have been noted to have small spiral structures around their cores that funnel in material, possibly to feed a supermassive black hole.

The Bubble Nebula

21 April 2016

A giant bubble measuring seven light years across is one strange sight Hubble has spotted on its universe safari. Also called NGC 7635, the shimmering sphere has been created as hot gas escaping from the massive star BD +60°2522 at over 6.4 million kilometres (4 million miles) per hour pushes against the colder interstellar gas. As the density of this gas isn't uniform, the bubble has been distorted around the star, which sits in the top left rather than the centre of the bubble. This star is 45 times more massive than the Sun and is aged at 4 million years old. It's expected to go out with a bang in the next 20 million years.

"The view of the Bubble Nebula, crafted from Wide Field Camera 3 images, reminds us that Hubble gives us a front-row seat to the awe-inspiring universe we live in"

John Grunsfeld, former NASA astronaut

Gravitational lensing

9 August 2021

This strange ring is the light from a distant quasar being warped around two massive foreground galaxies, with these giant ellipticals acting like a magnifying lens due to a line-of-sight effect. As the quasar's light travels towards us, it passes through space-time that's curved by the influence of the massive galaxies. This bending of the light magnifies it, allowing us to see much further into the universe if a foreground object is perfectly aligned. As the foreground object in this case is a pair of galaxies, light from the quasar 2M1310-1714 is also faintly visible in the centre – an even rarer phenomenon than gravitational lensing itself.

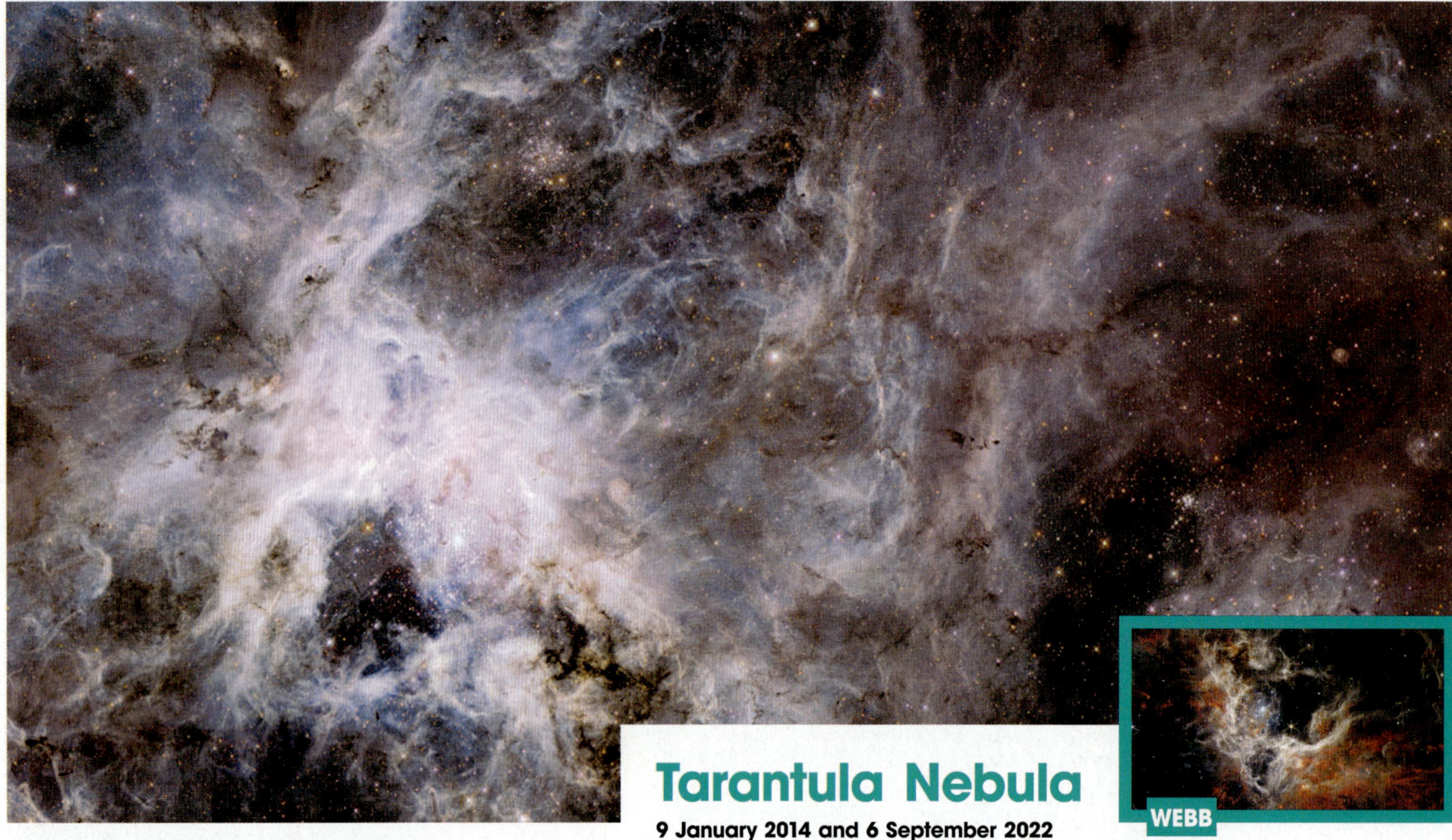

Tarantula Nebula

WEBB

9 January 2014 and 6 September 2022

The Tarantula Nebula is the most active starburst region in our galactic neighbourhood, bringing new stars to life in the Large Magellanic Cloud. This image was put together using data from the Hubble Tarantula Treasury Project (HTTP), an effort to map out and study the region's stellar population in order to surmise its overall structure. Webb's advanced eyes were able to reveal even more detail in the swirling wisps of dust and gas, as well as uncovering thousands of stars and background galaxies unseen by Hubble. This region is a favourite for studying stellar evolution, hosting so many stars of different ages and sizes.

Stephan's Quintet

9 September 2009 and 12 July 2022

This grouping of galaxies is formed from four interacting galaxies and one interloper – the spiral NGC 7320 in the top left, which is 250 million light years closer to us in the same line of sight. Far beyond this forefront galaxy, a distorted spiral and elliptical sit either side of a chaotic merger. Hubble spied this cosmic collection in 2009, revealing that gravitational interactions are causing mass star formation. Over a decade later, Webb observations have given scientists a new view, resolving many more stars and clusters and illuminating the cores of the galaxies.

WEBB

A worthy successor

The James Webb Space Telescope is by no means a replacement for Hubble

© Adrian Mann

After several delays, Webb was launched into space on 25 December 2021 and made the 1.5-million-kilometre (930,000-mile) journey to its orbital outpost at Lagrange point L2 before opening its eyes to the universe. Conceptualised as the Next Generation Space Telescope in 1996 – when Hubble was still in its infancy – it was never intended to usurp its predecessor, but instead achieve what the original space telescope couldn't. While Hubble does possess some infrared capabilities, its instruments weren't designed for these types of observations, focusing on visible and ultraviolet wavelengths. Webb was created to look beyond Hubble in the infrared, allowing it to view more distant objects with higher redshift. Though both telescopes have a similar purpose, each is able to see the universe in a different light, bringing more of the electromagnetic spectrum into view.

Pillars of Creation

5 January 2015 and 19 October 2022

Possibly the most iconic Hubble target, this star-forming section of the Eagle Nebula shot to fame in a 1995 image release, making Hubble a household name. For its 25th anniversary, Hubble turned back to the imposing pillars of gas and dust, which are filled with starbirth, using its upgraded Wide Field Camera 3 to get a higher resolution view. It seemed fitting to turn Webb towards the same structure, revealing a much more star-studded view of the region. Infrared observations can peer through cosmic dust and gas to reveal what's hidden behind the haze, bringing thousands of stars into view.

When stars go supernova

All about the titanic stellar eruptions that give rise to life

Place your fingers on your wrist and check your pulse. You can feel the blood pumping around your body, delivering oxygen from your head to your toes. That oxygen is ferried around by trillions of red blood cells, each just 0.008 millimetres across. The element iron plays an indispensable role in this oxygen delivery, but the universe would contain very little iron if it weren't for a ferocious type of exploding dying star called a supernova. Without supernovae, you simply wouldn't exist.

We often think of the Sun as the quintessential star, but it's not massive enough to go supernova when it dies. According to Dr Christopher Frohmaier, a supernova researcher at the University of Southampton, to detonate in this most spectacular of celestial firework displays, a star needs a starting mass equivalent to at least eight Suns. The path that these massive stars follow towards their eventual demise as a supernova is inevitable. It's triggered by a fundamental shift in the interplay between the outward pressure generated by the nuclear fusion reactions in the star's core and gravity. "Throughout a massive star's life, it's in a constant balance between these forces," says Frohmaier.

Things begin to change when the hydrogen that sustains the nuclear reactions runs out. "The star will burn sequentially heavier and heavier elements to support its outer layers," Frohmaier says. Nested shells of helium, carbon, oxygen, neon, magnesium and silicon build up as the star exhausts its supply of successive elements and desperately scrambles for something new to burn. Eventually, the star's nuclear reactions produce iron from the silicon, which builds up the middle of the core. "The phase of converting silicon to iron can be over in a matter of hours," Frohmaier says. That compares to perhaps billions of years burning hydrogen into helium.

It's at this point that the nuclear reactions switch off. Iron is the most stable element in the periodic table – burning iron requires more energy to go in than comes out, and so it simply doesn't happen. Now unsupported, the star's outer layers come crashing down, building up an intense pressure on the core that breaks the iron back down into helium. "It's called photodisintegration," Frohmaier says. "And it also creates an overwhelmingly large number of neutrinos."

Neutrinos are tiny, almost-massless subatomic particles. The photodisintegration of the iron core creates approximately 10 billion trillion trillion trillion trillion neutrinos. "The neutrinos are blown outwards and drive the supernova explosion by pushing against the star's outer layers," Frohmaier explains. It's a slightly surreal role for neutrinos to play, as they are famously very antisocial

A Type Ia supernova shining as brightly as the centre of the galaxy it's in

The star Wolf-Rayet 124 is giving hints that it's due to explode soon

© NASA, ESA, CSA, STScI, Webb ERO Production Team

particles. You would need a light year of lead to have a 50:50 chance of halting a single neutrino, yet there are so many produced by photodisintegration that their collective power blasts an entire massive star apart. In a short ten-second salvo, the neutrinos release almost a quadrillion quadrillion quadrillion Joules of energy. That's about the same amount of energy as the Sun will release in its entire 10-billion-year lifetime.

Astronomers call supernovae like these core-collapse supernovae. However, despite the huge amount of energy they release, when astronomers spot a core-collapse supernova in the night sky, they aren't observing the light from the explosion itself. "What we are seeing is radioactive decay," Frohmaier says. The supernova produces vast amounts of a rare form of the element nickel, known as nickel-56. It's unstable, and so breaks down - or decays - into cobalt and iron. This process creates a lot of gamma rays – the most energetic form of light. "All the material thrown out from the explosion gets energised by those gamma rays," says Frohmaier. That material then glows with visible light, which is what lights up the night sky and alerts astronomers to the death of a massive star.

And that light can be searingly bright. In 1054, Chinese astronomers wrote about a 'guest star' that suddenly appeared in the sky. It was so bright that it could be seen during the day for a month and took almost two years to fade from the night sky. Today we know that they witnessed a supernova explosion, the remnants of which form the Crab Nebula (Messier 1) in the constellation of Taurus, the Bull. Betelgeuse, not too far away in the constellation of Orion, will do something similar in the future. Astronomers can get deeper insights into the mechanics of supernovae by breaking up their light into spectra, much like how a raindrop splits sunlight to create a rainbow. Hidden within this stellar spectrum are so-called absorption features that point to

© ESO/L. Calçada

SN 1987A

Supernovae are reasonably common - about two explode every century in a galaxy like our Milky Way. With hundreds of billions of galaxies to observe, astronomers see them on a regular basis. Usually, they're far away, meaning astronomers can struggle to see them in detail. Astronomers were handed an astronomical gift in 1987 when a Type II supernova exploded in the Large Magellanic Cloud, one of the satellite galaxies of the Milky Way. At just 168,000 light years away, they could observe never-before-seen features of the explosion. It was the first time that astronomers were able to confirm that radioactive decay is responsible for the visible light we see from core-collapse supernovae, for example. SN 1987A was also pivotal in our understanding of the role that neutrinos play. Several hours before the visible light from the supernova arrived on Earth, three separate neutrino observatories pinged with neutrino detections. It confirmed predictions from theoretical models that 99 per cent of the energy of a supernova is carried away by these ghost-like subatomic particles.

How do stars detonate?

Stunning supernova explosions can come from very different stellar systems

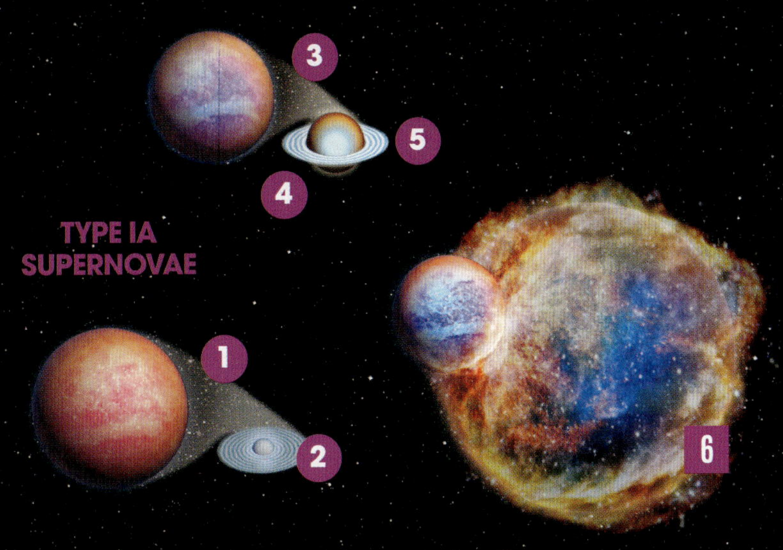

1 Finding the right stars
These supernovae occur when a white dwarf is found in a binary pair with another star.

2 Matter transfer
Two stars orbit close enough to each other that matter will transfer from the companion onto a white dwarf star in a process called accretion.

3 Squeezing down
This accretion of matter from its companion adds extra mass. The additional gravitational forces compress the already highly pressurised core of the white dwarf.

4 A star reborn
As the star approaches the Chandrasekhar limit, the dying star comes to life again and undergoes fusion in its centre.

5 Limited gains
However, white dwarfs can only get so massive through accretion. The Chandrasekhar limit puts their maximum size at about 1.4 times the mass of our Sun.

6 Outward pressure
This time the star cannot expand to accommodate the outward pressure. Instead, expansive forces build up.

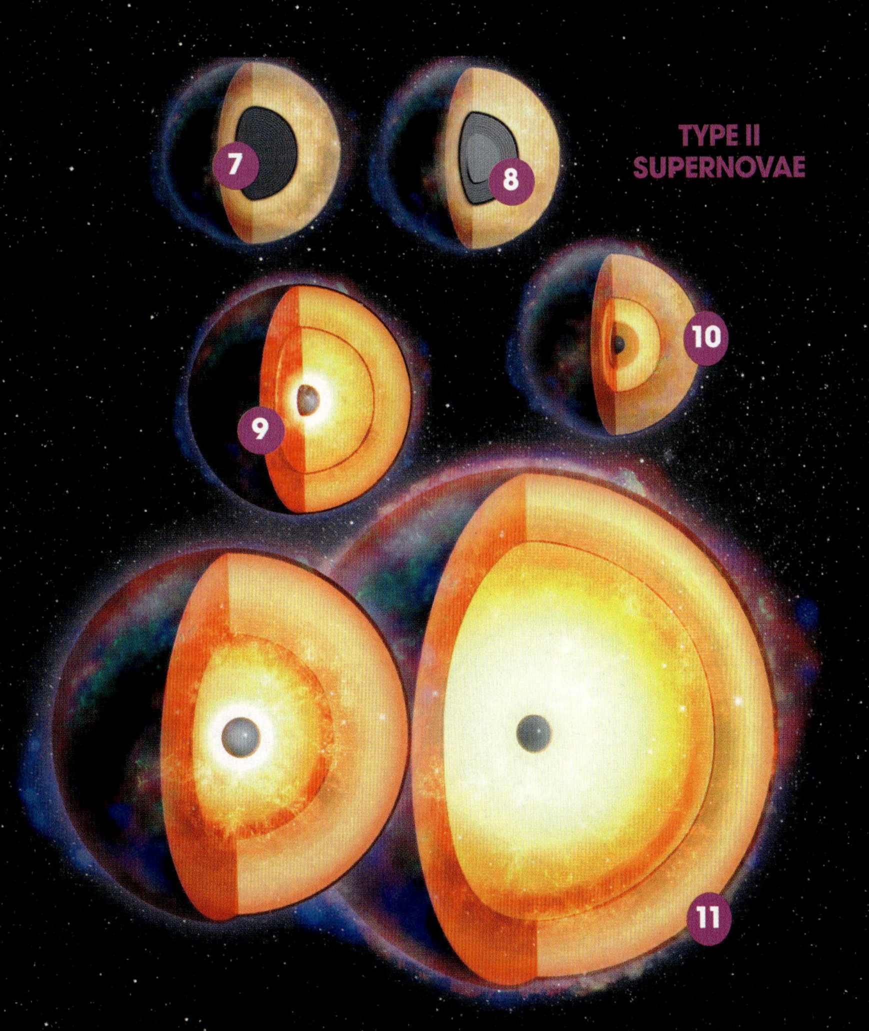

7 Out of gas
Supergiants like Betelgeuse will eventually run out of hydrogen and helium fuel, the nuclear burning of which prevents them collapsing in on themselves.

8 Elemental onion
Gradually, heavier elements build up at the star's centre, which becomes layered like an onion, with lighter elements towards the outside of the star.

9 Burning fuels
For stars this size, their mass and compressional forces at the core allow them to fuse heavier elements to maintain the outward pressure that fights gravity.

10 Resistance is futile
This act of resistance holds until the star's innermost core is composed of iron, which refuses to fuse. Gravity finally wins out, and the star's outer layers collapse inwards.

11 Shock wave rebound
This inward implosion bounces back off the core as a rebounding shock wave that blows the whole atmospheric envelope off the star, creating the spectacular supernova.

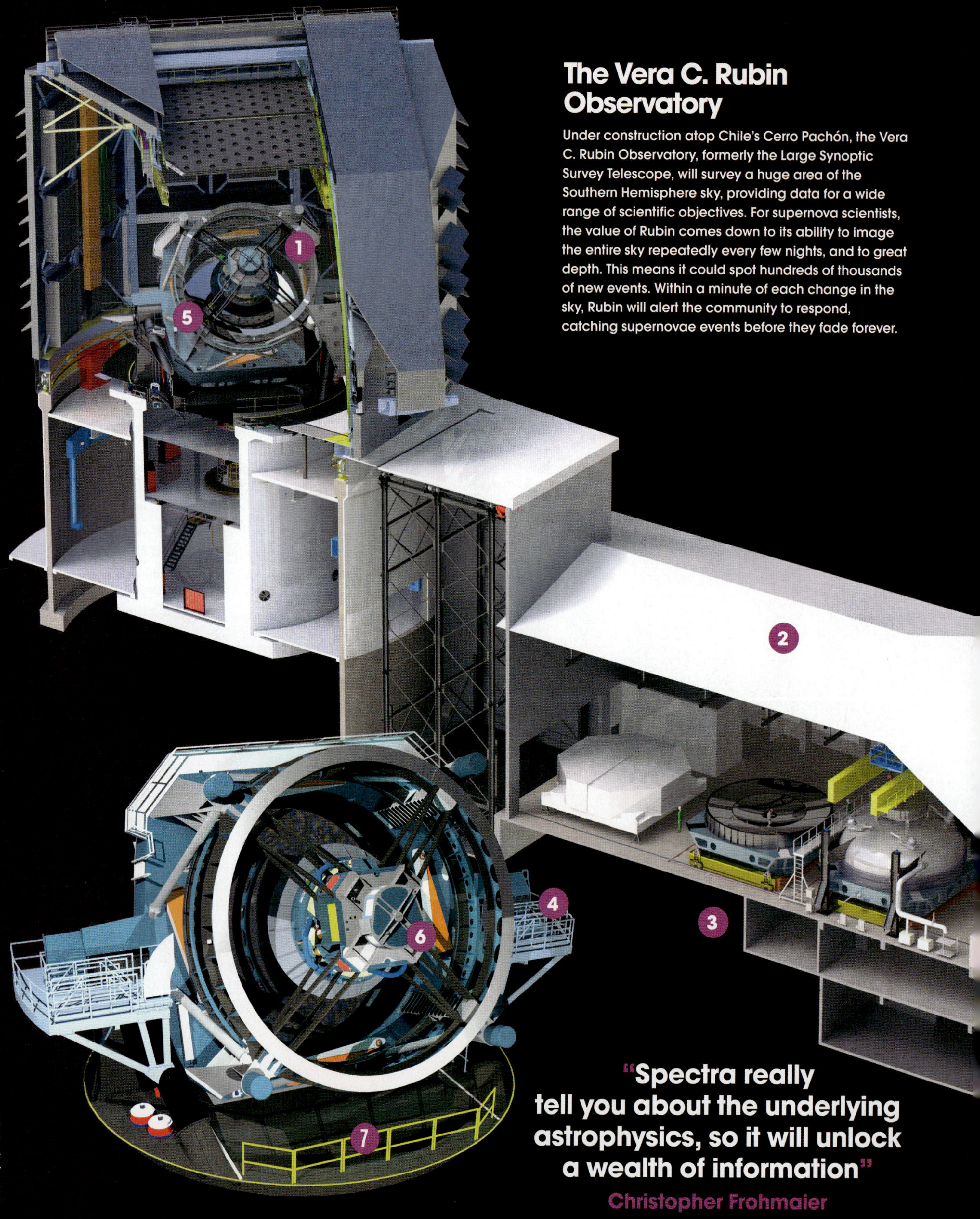

The Vera C. Rubin Observatory

Under construction atop Chile's Cerro Pachón, the Vera C. Rubin Observatory, formerly the Large Synoptic Survey Telescope, will survey a huge area of the Southern Hemisphere sky, providing data for a wide range of scientific objectives. For supernova scientists, the value of Rubin comes down to its ability to image the entire sky repeatedly every few nights, and to great depth. This means it could spot hundreds of thousands of new events. Within a minute of each change in the sky, Rubin will alert the community to respond, catching supernovae events before they fade forever.

"Spectra really tell you about the underlying astrophysics, so it will unlock a wealth of information"

Christopher Frohmaier

1 Large mirror
Rubin's reflecting telescope contains an 8.4-metre (28-foot) primary mirror that can track across at sufficient pace to image the entire sky every few nights.

2 Environment
The design of the observatory takes advantage of the mountain's natural topography. Its orientation was selected after extensive weather testing.

3 Keeping cool
To protect the mirror, the heated operation spaces are located below the service level. The heat-generating equipment is below that, farthest from the telescope.

4 Wide-angle lens
To maximise its potential to spot new transient phenomena, Rubin has a 3.5-degree field of view – the Sun as seen from Earth is only 0.5 degrees across.

5 Multi-band
Rubin will carry out a deep ten-year imaging survey in six broad optical bands to maximise the data provided on any supernovae or progenitors imaged.

6 On-site maintenance
The observatory has its own dedicated cleaning and coating area where the mirrors are washed and recoated during operations.

7 Big camera
Images will be recorded by a 3.2-gigapixel camera. It will take a 15-second exposure every 20 seconds to help detect new high-energy events.

The lens of the camera that will conduct the LSST

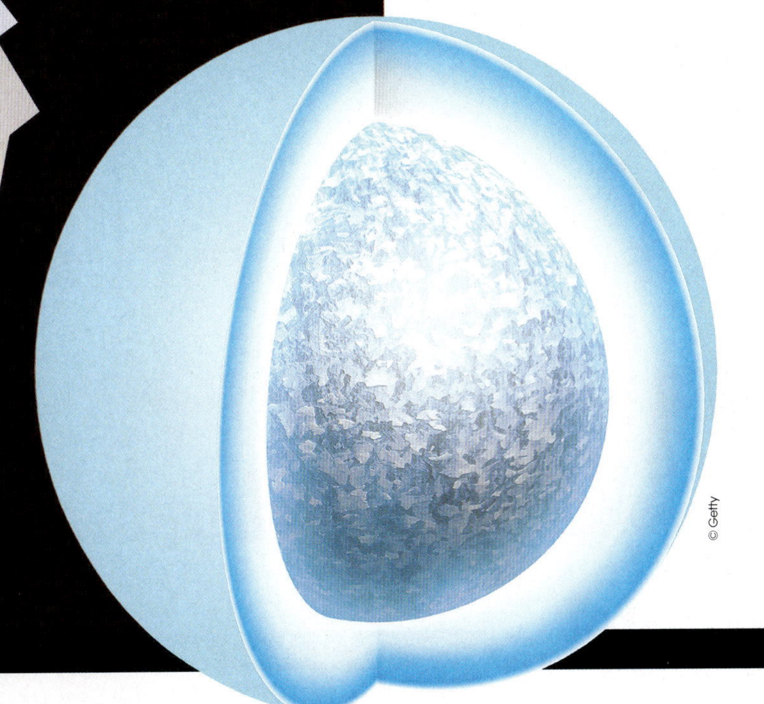

About the size of Earth, a white dwarf is made of carbon and oxygen

the presence of various elements within the debris from the supernova. All supernovae with hydrogen in their spectra are classed as Type II supernovae.

There's a separate class of supernovae, known as Type I, that don't have hydrogen in their spectra. For Type Ib, that's because fierce winds blew the hydrogen outwards from the star before it collapsed. For Type Ic the star also blew its helium away, meaning there's no hydrogen or helium present in the spectrum. These are still core-collapse supernovae, though. But by far the most famous Type I supernova – known as Type Ia – has a completely different origin. "No one knows exactly what the explosion mechanism is," Frohmaier says, although all options point towards another remnant of dead stars: white dwarfs.

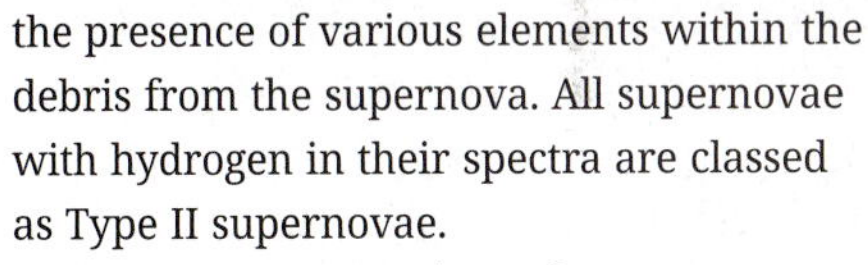

Stars that start off less massive than eight Suns aren't big enough to trigger core-collapse supernovae. These stars move from burning hydrogen to burning helium, eventually fashioning a core of carbon and oxygen about the same size as Earth... a white dwarf. There isn't sufficient pressure to fuse carbon, so why doesn't a white dwarf collapse? "White dwarfs aren't supported by fusion, but by something called degeneracy pressure," Frohmaier says. There's a rule in quantum physics that says it's impossible to force certain particles into the same state, so try as it might to collapse, a white dwarf cannot.

There's a limit to how massive a white dwarf can be, however. "If you exert too much pressure then the temperature will rapidly increase and the carbon will undergo thermonuclear runaway," Frohmaier says. In other words, it detonates like a nuclear bomb and explodes outwards as a Type Ia supernova. This mass threshold is known as the Chandrasekhar limit after the Indian astrophysicist Subrahmanyan Chandrasekhar, who calculated it aged just 19 during a sea voyage to Europe in 1930. It's possible for a stable white dwarf to approach this threshold by gaining mass from somewhere else.

Frohmaier says there are two main options being considered by astrophysicists. The first, called the single-degenerate scenario, involves the white dwarf ripping gas from a companion star. This could be an ordinary star like the Sun, or a Sun-like star that's itself on the way out, having puffed up into a so-called red giant. However, the ensuing supernova would interact with some of the hydrogen of the companion star, causing it to glow. "No one has seen that hydrogen," says Frohmaier, which means that most researchers are starting to favour the so-called double-degenerate scenario instead. Picture two Sun-like stars orbiting one another in a binary system. Such a scenario is very common, as at least half of the stars in the universe are thought to exist in these binary pairs. Both stars reach the end of their lives and become white dwarfs. The two white dwarfs circle around one another, gradually spiralling inwards towards oblivion. "They collide into a single white dwarf that's too massive to exist," Frohmaier says, and in a matter of milliseconds they detonate as a Type Ia supernova.

"Two white dwarfs circle around one another, gradually spiralling inwards towards oblivion"

Understanding what causes these enigmatic explosions is vital because Type I supernovae are used as cosmic rulers to measure distances in the universe. If every Type Ia supernova is the result of a detonation of a white dwarf, or dwarfs, with a total mass close to the Chandrasekhar limit, then each explosion will have a similar brightness, thus those that appear dimmer to us must be further away. In the late 1990s, astronomers used measurements of Type Ia supernovae to show that the expansion of the universe appears to be speeding up. If it continues to accelerate as it is, then all structures in the universe will be torn apart in around 22 billion years in an event astronomers call the 'Big Rip'. Exactly what's driving the quickening expansion is unclear, but cosmologists point to the influence of a mysterious entity dubbed dark energy – yet they don't know what it is or how it works. Understanding Type Ia supernovae better might provide valuable clues.

Astronomers refer to objects like Type I supernovae that have an inherently fixed brightness and can be used to measure cosmic distances as standard candles.

A supernova would have a devastating impact on planets and its wider solar system

Five stars ready to explode

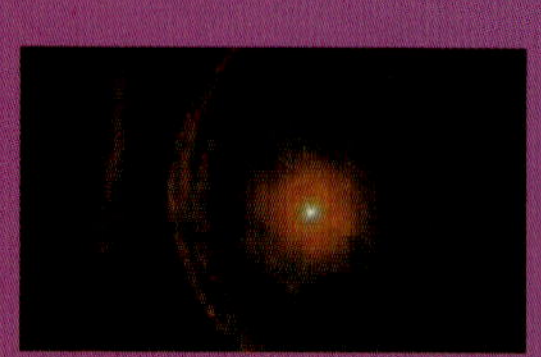

Betelgeuse

Stellar evolution models suggest it should go supernova anytime in the next 100,000 years, when it will be as bright as the full Moon in the night sky. Even before its recent dimming, Betelgeuse showed the most variation of any progenitors.

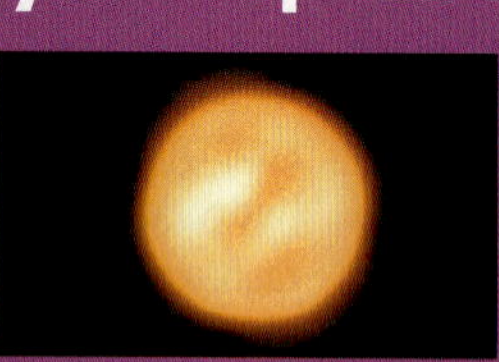

Antares

The red eye of the scorpion in the constellation of Scorpius is due to go supernova in the next million years. With a mass 15 to 18 times that of our Sun, it will be spectacular, though its lack of variability suggests it's not quite ready yet.

Mu Cephei

Known as 'Herschel's Garnet Star', it's further away than Betelgeuse, but double its diameter, and therefore perhaps closer to death. Its supernova would likely produce a black hole while providing a 'guest star' as bright as Venus.

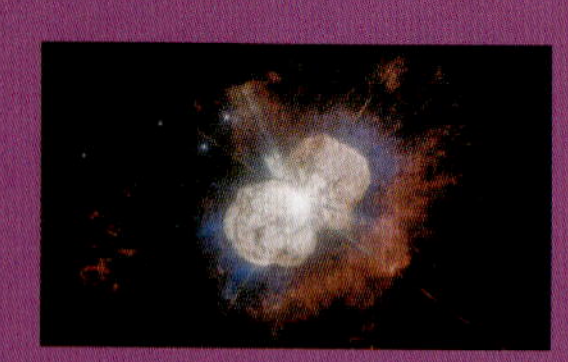

Eta Carinae

Observers have already seen this binary star throw out its outermost layers in a series of pulses. Much further away than Betelgeuse, its impending death could produce a gamma-ray burst or a superluminous supernova, easily visible from Earth.

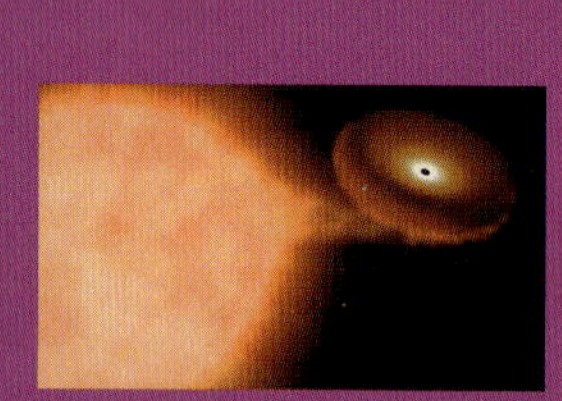

V Sagittae

Made up of an ordinary star spiralling in towards a white dwarf, researchers have predicted a date later in this century for a supernova event that will create a 'new star' as bright as Sirius in the night sky.

© Tobias Roetsch

Frohmaier prefers to call Type Ia supernovae 'standardisable candles'. "There's some variation in the brightness of these objects," he says. Astronomers have to make corrections for these subtle differences in order to use them as celestial rulers. "That's worked well for 20 years, but as we enter the next era of telescopes we're being held back by our lack of understanding of astrophysics."

At the time of writing (March 2023), astronomers may not have to wait much longer for answers. Construction is almost finished on the much anticipated Vera C. Rubin Observatory in Chile, which should see the start of full operation towards the end of 2024. It's home to the largest digital camera ever constructed, with a total of 3,200 megapixels, and it's about the size of a person. This colossal camera will be at the heart of an observing run known as the Legacy Survey of Space and Time (LSST). Astronomers expect to spot hundreds of thousands of new supernovae. "It will be revolutionary for cosmology," Frohmaier says.

While this will spot supernovae, it won't be able to measure their spectra. For that astronomers will turn to another facility in Chile. The 4-metre Multi-Object Spectrograph Telescope (4MOST) is an instrument attached to the Visible and Infrared Survey Telescope for Astronomy (VISTA) at Paranal Observatory. Rather than measuring the spectrum of one supernova at a time, 4MOST will be able to observe the spectra of 2,400 supernovae simultaneously. "Spectra really tell you about the underlying astrophysics, so it will unlock a wealth of information," says Frohmaier, who was announced as the project's deputy project scientist. It will allow astronomers to compare models of supernova explosions built in supercomputers to real data to check if they've built a realistic model.

Perhaps astronomers will finally be able to tease apart the mysteries that surround Type Ia supernova, not only measuring distances in space more accurately, but also getting a clearer understanding of why our universe continues to expand at an ever quickening pace.

Colin Stuart
Astronomer and space science writer
Colin holds a degree in astrophysics, has written over 17 books on space and has an asteroid named in his honour.

Dark energy: The force tearing space apart

It's the most mystifying phenomenon in the universe, but we're hot on its trail

Our universe is growing. Ever since the Big Bang, every point in the fabric of space has been expanding in all directions. This expansion is carrying almost all galaxies away from us. The biggest surprise came in 1998 with the discovery that not only is the universe expanding, but that expansion is accelerating. Nobody knows why, but scientists have come up with a term for the mysterious force driving the acceleration: dark energy.

According to data from the European Space Agency's (ESA) Planck spacecraft, dark energy constitutes over two-thirds of all the mass and energy in the universe, or 68.3 per cent. The remainder is 26.8 per cent dark matter and just 4.9 per cent normal matter, which makes up the stars, galaxies and planets. Scientists are perplexed as to what dark energy is. There are ideas, but nothing concrete. It's important to try and figure out the puzzle of dark energy because the fate of the universe depends on it. Prior to the discovery of dark energy, cosmologists had expected to find that the expansion of the universe was running out of steam 13.8 billion years after the Big Bang. If that was the case, the universe could have gone in three directions depending upon how much matter, and therefore gravity, there was in the universe.

If there was enough matter, its gravity would act on the expanding universe, slowing the expansion and gradually overcoming it, eventually causing the universe to begin to shrink again before collapsing in a 'Big Crunch'. If the amount of matter and gravity were finely balanced with the energy of the expanding universe, it would create a static universe that would remain forever. However, if there was not enough matter in the universe to counteract the expansion, then the universe would continue to expand forever, taking all the galaxies with it until they disappeared over the cosmic horizon, leaving the Milky Way all alone.

With the discovery of dark energy, the fate of the universe now seems much clearer. If dark energy keeps its strength and continues accelerating the universe's expansion, then it's more likely to expand forever. In the worst-case scenario, the expansion could be so extreme that it begins to tear galaxies, stars, planets and even atoms apart. However, since we don't know what dark energy is, we cannot predict what it is going to do in the future. Astronomers can measure what it is doing today and what it has done in the past, however, and are able to make some extrapolations based on that.

Dark energy was discovered by looking at the light of exploding stars in faraway galaxies. In particular, astronomers were looking at a breed of supernovae known as Type Ia. These are the explosions of white dwarf stars, and they tend to all explode with the same peak brightness. From our viewpoint, millions if not billions of light years away from these supernovae, they appear pretty faint. But because we know how bright they would be if we were up close to them, we can calculate how far away they are. Astronomers call these 'standard candles' and use them to measure distances across the universe. These distances are then compared to their cosmological redshift, which is the amount their light is stretched into redder wavelengths by the expansion of the universe.

An image of a galaxy cluster, with hot intracluster gas shown in pink. The blue overlay reveals the calculated areas of dark matter

What is dark energy?

All About Space asks the astrophysicists

"Dark energy makes up about 70 per cent of the universe. It seems to be evenly spread throughout. What we have discovered about dark energy is that it 'pushes', as in it repels outwards. It makes the entire universe – which is already expanding – expand faster."

Dr Karl Kruszelnicki, University of Sydney

"Dark energy is incredibly strange, but it makes sense that it went unnoticed. I have no clue what dark energy is. It appears strong enough to push the entire universe, yet its source is unknown, its location is unknown and its physics are highly speculative."

Dr Adam Riess, Johns Hopkins University

"It could pull the universe apart. If there's sufficient dark energy, the universe will at some point pull itself apart completely in a 'Big Rip'. The only problem is that we have no clue what dark energy is."

Dr Hitoshi Murayama, University of California, Berkeley

RS Puppis is an example of a Cepheid variable. These were essential in showing an expanding universe due to their luminosity-pulsation period relationship

It was the fact that the distances to these supernovae were greater than their redshift that implied that the expansion was getting faster, not slower, as identified by two teams of scientists: the Supernova Cosmology Project, led by Saul Perlmutter of Lawrence Berkeley National Laboratory, and the High-Z Supernova Search Team (Z is astronomy shorthand for redshift) led by Harvard University's Brian Schmidt and Adam Riess of Johns Hopkins University.

Riess led a team of scientists who used Type Ia supernovae to discover that the expansion of the universe is even faster than previously thought. They searched for supernovae in galaxies in which they could also see a special type of variable star called Cepheid variables. These pulsating stars are also standard candles as their maximum luminosity depends on the length of their variation – the longer it takes for them to reach peak brightness and then fade back down to normal levels as they pulse, the brighter their peak will be.

Comparing the Cepheid data with the distance of those galaxies as measured by Type Ia supernovae, Riess' team was able to calibrate the supernova distance markers, allowing them to more accurately measure the distances to even more remote galaxies. While the Planck spacecraft had measured the expansion rate of the universe as being 66.5 kilometres (41.3 miles) per second per megaparsec – equal to 3.26 million light years – Riess' team found that actually the universe was expanding much faster, at 73.2 kilometres (45.5 miles) per second per megaparsec. That doesn't mean that Planck made a mistake. The rate of expansion that it measured – known as the Hubble constant –

"We may not have the right understanding, and it changes how big the Hubble constant should be today"
Adam Riess

The Dark Energy Survey, an international effort based at the Cerro Tololo Inter-American Observatory, Chile, began searching the southern skies on 31 August 2013

Dark energy vs dark matter

Dark matter

It's entirely invisible and neither emits or absorbs heat, light or radiation of any kind.

It can be detected indirectly by its gravitational effect on normal matter and space-time.

It comprises 84.54 per cent of all matter in the universe, or 26.8 per cent of the universe's total make-up.

It plays a crucial role in galaxy formation and stops galaxies from flying apart.

Galaxies are pulled together by dark matter

The Big Bang, 13.8 billion years ago

Universe is pushed apart by dark energy

Dark energy

It's accelerating the expansion of space, shown by studies of supernovae in distant galaxies.

It exerts a small negative pressure that's constant throughout space and acts against gravity.

It affects the shape of the universe and its large-scale structure, such as how galaxies are spread out.

It was first discovered by astronomers in the late 1990s.

Dark energy mappers

A new generation of instruments promises to hunt down dark energy, exposing its secrets

DOE/FNAL/DECam/R. Hahn/CTIO/NOIRLab/NSF/AURA

1 Dark Energy Survey

This was a visible and near-infrared survey of the universe using the four-metre (13-foot) Blanco Telescope at the Cerro Tololo Inter-American Observatory in Chile. The telescope was fitted with the state-of-the-art robotic Dark Energy Camera (DECam), which surveyed 300 million extremely faint galaxies and thousands of supernovae over an eighth of the total sky. The DECam instrument made repeated observations of certain areas of the sky in various wavelengths, as well as more long-period observations to pick out the faintest galaxies. The project's results hope to discover if dark energy's density changes over time.

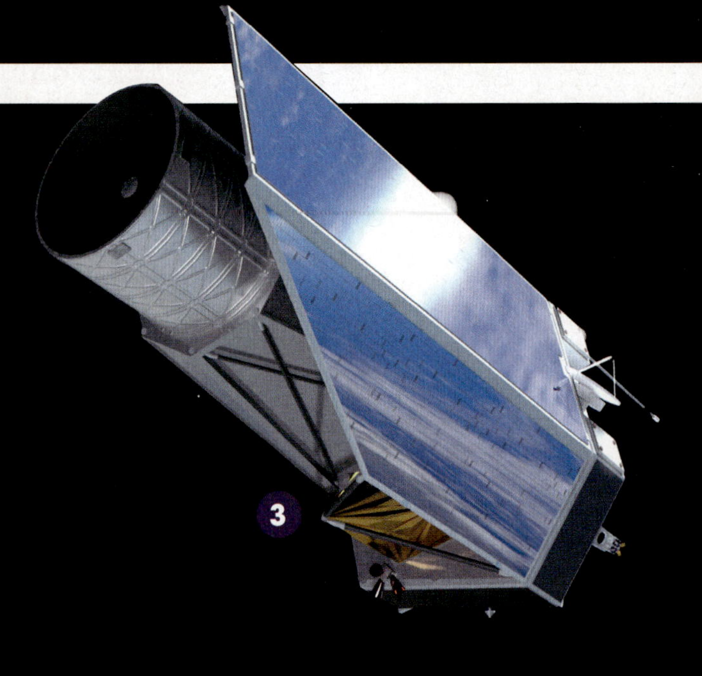

3 Euclid space telescope

How has dark energy contributed to the universe's acceleration over cosmic time? This is what the Euclid spacecraft will try to ascertain when it's launched in 2023. It will do this by measuring the redshifts of galaxies back to when the universe was just 28 per cent of its current age. It will also look at gravitational weak lensing as well as ripples in normal matter – usually in hydrogen gas between galaxies. To get decent results, Euclid will survey at least half the entire sky with its visible and near-infrared cameras.

> **"Galaxy clusters are like Russian dolls, with smaller ones having a similar shape to the larger ones"**
> Andrea Morandi

2 Square Kilometre Array

The Square Kilometre Array (SKA) will span two continents when it begins operation in 2027. As the largest radio telescope network in history, all of the SKA's individual elements will be able to act like a single super-continental radio dish, giving it the ability to study the universe in resolutions never before seen in radio wavelengths. In particular, emissions from hydrogen gas – the most abundant element in space – will be mapped in three dimensions from the distant past to the present day. The SKA's high resolving power could reveal dark energy's effects from ripples in the gas and more information on galactic evolution.

An extraordinary example of gravitational lensing in galaxy cluster Abell 2218 due to dark matter. There's not enough visible matter to distort the light from background galaxies

was from the very beginning of the universe, observed in the light of the cosmic microwave background (CMB) radiation just 380,000 years after the Big Bang.

On the other hand, Riess' team measured the Hubble constant in the modern universe. It isn't clear why the two values are different, but one explanation could be that dark energy has grown stronger over the years, speeding up the expansion, meaning that the Hubble constant isn't actually constant at all. If that's true, it has serious implications for the fate of the universe.

"If we know the initial amounts of stuff in the universe, such as dark energy and dark matter, and we have the physics correct, then you can go from a measurement at the time shortly after the Big Bang and use that to predict how fast the universe should be expanding today," says Riess. "However, if this discrepancy holds up it appears we may not have the right understanding, and it changes how big the Hubble constant should be today." Riess' findings also somewhat contradict another dark energy discovery made by scientists who used NASA's Chandra X-ray Observatory to observe galaxy clusters. Because galaxy clusters are filled with gas as hot as 1 million degrees Celsius (1.8 million degrees Fahrenheit) or more, they glow brightly in X-rays. These X-ray-emitting regions of galaxy clusters all have similar shapes and sizes relative to the size of the clusters themselves.

"In this sense, galaxy clusters are like Russian dolls, with smaller ones having a similar shape to the larger ones," says Andrea Morandi of the University of Alabama in Huntsville, who led the team. "Knowing this lets us compare them and accurately determine their distances across billions of light years." From these distances, the true size of the clusters can be gauged. Because the growth of galaxy clusters is ultimately stunted by dark energy pulling away galaxies that would otherwise join the clusters, how big galaxy clusters can reach during different epochs of history is an indication of how strong dark energy is and how it has acted over billions of years. Since a galaxy cluster's X-ray profile is a proxy for its true physical size, by measuring them with Chandra, Morandi's team found that dark energy

Riess and his High-Z Supernova Team published some of the first evidence that the expansion of the universe is accelerating

Evidence for dark energy

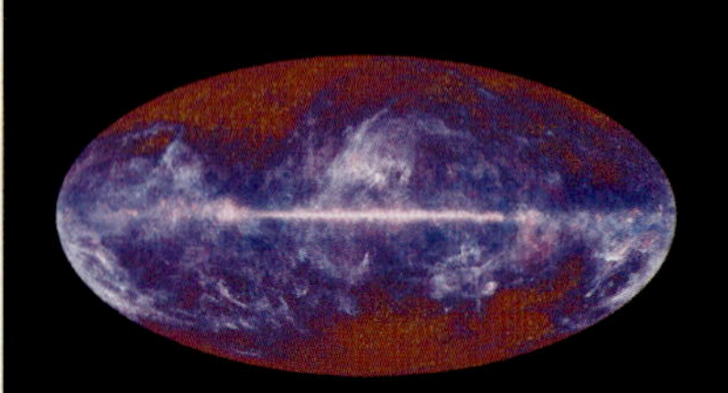

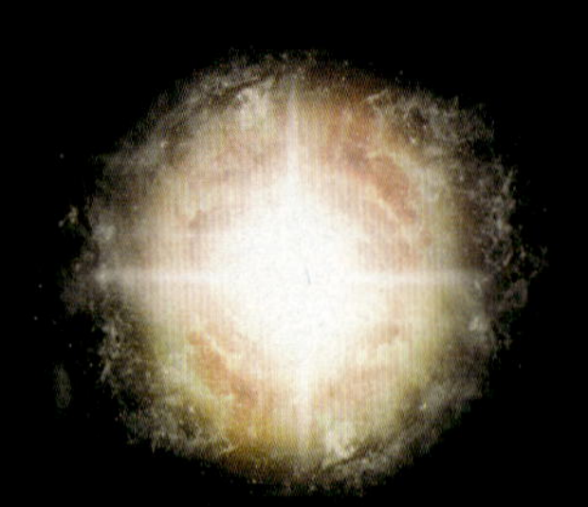

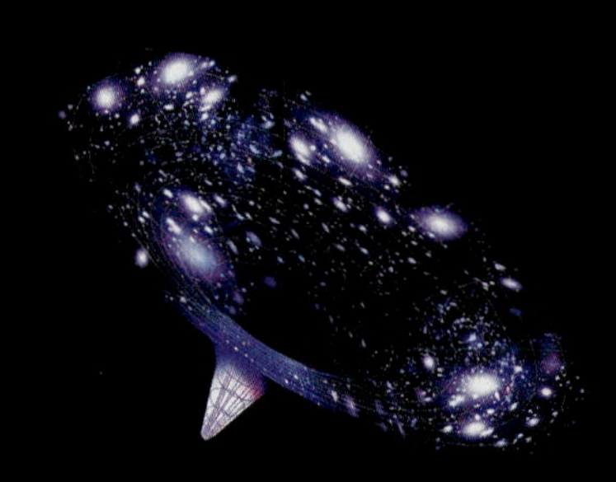

Relic radiation

The CMB is the primordial 'echo' of the universe. Measurements taken by probes show that the universe's geometry is nearly flat. For that to be the case, the total amount of matter in space must tally with the average density of matter to energy in the cosmos. But the CMB's spectrum shows that this isn't what's happening – matter only accounts for 30 per cent of the density.

Exploding stars

Type Ia supernovae occur in binary systems when a white dwarf's gravity pulls material from the other star onto itself, causing it to explode, briefly outshining all other stars in its galaxy. Their intrinsic brightness is known, so their distances can be calculated from how dim they appear. Using this, scientists were able to show that the universe's expansion is accelerating.

The shape of the cosmos

Its shape is governed by the general theory of relativity, which describes how space-time is warped by matter. Measurements of the CMB show that the universe is nearly flat, but there's not enough matter to keep it that way. Since E=mc2 shows that matter and energy are two sides of the same coin, could dark energy remedy the shortfall?

The universe is expanding

The expansion rate of the universe is the time it takes to expand a certain amount. In a certain period of time, space will expand by a certain amount – whatever makes space expand doesn't dilute or decrease in density. An expanding universe leads to a greater amount of space, and more of what caused the expansion. This suggests dark energy may be a feature of space itself.

doesn't seem to have become stronger or weaker in the last 8.7 billion years. This might mean that a change in the expansion rate of the universe occurred before then, closer to the time of the Big Bang.

Galaxies can also reveal the presence of dark energy in other ways. We've seen standard candles in the form of supernovae, so now meet 'standard rulers', physical yardsticks that we can measure dark energy against. The distribution of galaxies in the universe can be traced back to perturbations seen in the cosmic microwave background, which resulted in a standard distance between galaxies. This standard distance can act like a ruler, and by measuring how the distance between galaxies has grown as the universe has expanded, astronomers can measure the strength of dark energy.

However, for standard candles and standard rulers to really make a difference in our quest to solve the mystery of dark energy, what scientists really need is more data. This data will come from giant galaxy surveys conducted by both general-purpose observatories and dedicated dark energy projects. First up was the Dark Energy Survey, an international project to detect many thousands of Type Ia supernovae which also aimed to map out the large-scale structure of the universe in the form of hundreds of millions of galaxies, providing many more examples of standard rulers. The Dark Energy Survey made observations from 2013 to 2019 using the four-metre (13-foot) Víctor M. Blanco Telescope and its powerful 570-megapixel Dark Energy Camera at the Cerro Tololo Inter-American Observatory, situated in Chile. The project released its first batch of data for scientists to analyse in January 2016.

"Thanks to the extreme sensitivity of the camera and to the large area of sky that can be imaged through the telescope at once – 15 times the size of the full Moon – we expect the Dark Energy Survey to find more supernovae than any previous experiment," comments Dr Chris D'Andrea of the University of Portsmouth. In the first few months alone, it found 200 Type Ia supernovae. However, we're still a few years away from having a thorough analysis of the results.

There's also the Dark Energy Spectroscopic Instrument, or DESI for short, at the four-metre (13-foot) Nicholas U. Mayall Telescope at

> **"The Euclid mission will map the size and location of a whopping 2 billion galaxies, more than have ever been mapped before"**

Kitt Peak National Observatory in Arizona. It will measure the cosmological redshifts of 30 million galaxies out to a distance of 10 billion light years in order to create a huge map of the large-scale structure in the universe. It will look at how the expansion of the universe has stretched the space between individual galaxies, as well as the size of galaxy clusters. It saw first light in 2019, just as the Dark Energy Survey completed its work, so we will have to wait a while for its results. It will build on the efforts of previous surveys such as BOSS, the Baryon Oscillation Spectroscopic Survey, which was part of the Sloan Digital Sky Survey based in New Mexico. This measured the redshifts of a million galaxies across 6 billion light years and allowed comparisons to be made of the growth of standard rulers between the modern day and 6 billion years ago.

It's an even longer time frame for other projects. The most powerful digital camera in the world, featuring 3.2 billion pixels, has been fitted to the heart of the Vera C. Rubin Observatory – a huge 8.4-metre (27.6-foot) telescope on a mountain called Cerro Pachón in Chile that is expected to begin survey operations in late 2024. While it will be a general-purpose survey telescope, its Legacy Survey of Space and Time (LSST) will map millions of galaxies and find many more supernovae to aid in the quest to learn more about dark energy.

But there are more ways to study dark energy than just with visible light. The Square Kilometre Array, a vast interconnected network of thousands of small radio telescopes that will come online in the next decade, will study dark energy in two ways. First, it will listen for how the gravity of galaxy clusters affects the passage of radio waves through the universe, with the size of the radio distortion being a signal to indicate the different sizes of standard rulers across billions of years. It will also complete a survey of galaxies via the 21-centimetre (8.3-inch) wavelength radio emissions from neutral hydrogen in those galaxies, adding to the many galaxies observed by other surveys to help create a comprehensive map of the universe from which we will be able to measure dark energy across space and time.

Finally, the quest for dark energy will be heading into space with the joint ESA and NASA Euclid mission, set to launch in July 2023. Armed with a 1.2-metre (3.9-foot) telescope – a decent size for a space mission – and a 600-megapixel digital camera, it will map the size and location of a whopping 2 billion galaxies, more than have ever been mapped before, across 10 billion light years. It will tell the story of the evolution of the standard rulers in the universe as well as highlight the result of the battles between gravity and dark energy in the growth of galaxy clusters.

"Euclid will provide a wealth of data on the three-dimensional matter distribution in the universe," says Ralf Bender of the Max Planck Institute for Extraterrestrial Physics in Germany, who is a scientist working on the Euclid mission. "Not only will this give us interesting insights into the evolution of galaxies and galaxy clusters, but we will also be able to better understand the accelerating expansion of the universe. Hopefully this will bring us a big step forward in solving the riddle that is dark energy."

Kulvinder Singh Chadha
Space science writer
Kulvinder is a freelance science writer, outreach worker and former assistant editor of *Astronomy Now*. He holds a degree in astrophysics.

The 64-dish MeerKAT array in Carnarvon, South Africa, will be integrated into the Square Kilometre Array

Why are we still searching for intelligent alien life?

Should we widen our search criteria when looking for extraterrestrials?

Humans have long scanned the heavens for signs of other advanced civilisations in the universe. And we've found nothing. Absolutely nothing. So maybe we shouldn't be so focused on intelligent life, but on any sort of life whatsoever. Sure, a tiny microbe may not be as exciting as swapping stories with distant aliens, but signs of non-intelligent life may be much more common – and much easier to find – in our galaxy.

Life, including intelligent life, evolved here on Earth. Yet there shouldn't be anything particularly remarkable about our planet; it's just another random world in the galaxy. So if intelligent life happened here, it must be pretty common – widespread enough that we should be seeing signs of alien civilisations all over the place. So, where is everybody? This is the heart of the infamous Fermi paradox, and it's the main argument used to fuel the search for extraterrestrial intelligence (SETI). At first glance, it seems immediately obvious that we should not be alone in the universe, and so if we look hard enough, we should see some evidence for intelligence.

Perhaps aliens are blasting radio signals for us to listen to. Perhaps they're just generally blasting the radio, and we happen to pick it up. Perhaps they've left artefacts in the Solar System, designed to monitor us or just hang around. Perhaps they'll engage in mega-engineering projects, like enclosing their star in a swarm of solar panels. Or perhaps they'll just play around and contaminate their star with heavy metals to announce their presence. After over half a century of SETI, however, we've found nothing. No radio signals. No artefacts. No mega-engineering. To date, after over a hundred dedicated searches, we have absolutely no evidence for any intelligent life in our galaxy, or even in the universe.

Expert: Paul M. Sutter

Sutter is a research professor in astrophysics at the Institute for Advanced Computational Science at Stony Brook University and a guest researcher at the Flatiron Institute in New York City. He is also the author of two books: *Your Place in the Universe* and *How to Die in Space*.

> "Signs of non-intelligent life may be much more common – and much easier to find – in our galaxy"

Should we lower our standards?

Perhaps we should just search for extraterrestrial life, rather than focusing on advanced alien civilisations. That means any kind of life: single-celled organisms floating in oceans, moss clinging to rocks or the first hints of complex creatures moving around their environments. Sure, these kinds of life forms may not be as loud as intelligent life, but that doesn't make them invisible. Indeed, one of the key features of any kind of life is the ability to throw a planet out of equilibrium.

First contact

Our current technology limits us to studying the atmospheres of giant planets orbiting close to their parent stars. But NASA's Transiting Exoplanet Survey Satellite (TESS) is cataloguing a number of promising candidates for follow-up studies with the James Webb Space Telescope, which will have the capability to detect an overabundance of oxygen in the atmospheres of alien worlds. As first-contact scenarios go, it may not be that exciting. It's likely our first evidence for life outside Earth will take the form of a wiggle in a line on a plot, telling us that living creatures have dramatically altered the equilibrium of their home planet.

SETI assumption

The assumption behind SETI is that intelligent life should be easier to detect than regular life, because intelligent creatures are capable of making their presence known. But something in this argument is going wrong. Either intelligent life isn't as common as we might have hoped, or it's not as detectable as we might have hoped.

Atmospheric clues

Our planet probably formed with a good deal of oxygen; there's plenty of it in the universe to go around. But oxygen is highly volatile and reactive, and it doesn't really last long on its own in an atmosphere – it either escapes into space or binds with other elements and turns into other things, like carbon dioxide or silicon dioxide. But around 2 billion years ago, a planet full of single-celled photosynthetic organisms ate enough carbon dioxide and burped out enough oxygen to completely revamp Earth's atmosphere, giving it substantially more oxygen than it would have in equilibrium. Life on Earth changed the very character of this planet's atmosphere. And that's detectable elsewhere.

Are we living in a multiverse?

Could our entire universe be just one small part of many? The evidence is mounting up...

Have you ever wondered how different your life would be if you had made different decisions? You may never know what may have been, but incredibly, all of those 'what if' scenarios may have played out somewhere. Over the past few years, support for the idea that we live in a 'multiverse', in which our universe is just one tiny bubble among countless others, has been gaining strength in recent years.

For many people, the term 'multiverse' conjures up pictures of parallel realities, some with just a slight difference from our own. And that sort of multiverse is indeed predicted by the 'many worlds' interpretation of quantum mechanics – the strange but highly successful model of how the universe works on the smallest scales – wherein every possible quantum state branches off into a new universe. In other words, every action that is physically possible, every choice that can take place, can and will happen, somewhere.

Such parallel universes might exist, and evidence for this 'many worlds' interpretation of quantum mechanics that invokes them might one day be found, but they would be forever unobservable, separated from our universe in ways we can hardly comprehend. Cosmologists like Matthew Kleban are instead interested in a more concrete form of multiverse – something beyond the realm of our current universe, but which we might nevertheless learn about. "We have a horizon in cosmology that's a lot like the horizon on Earth," explains Kleban, a professor of physics at New York University and a leading multiverse theorist. "If you're on an island in the ocean and climb to the highest point, there's a finite distance you can see, and you don't know what's beyond that horizon by directly seeing it. But you still might be able to get information about it – say a log comes floating up to your island with some plants growing on it. You can learn things from over the horizon because signals of various sorts can reach you from beyond it."

In the case of our universe, the horizon is a lot further away than a horizon on Earth – it's about 46.5 billion light years away from us in every direction. This apparent 'edge' to the universe is caused by the limited speed of light and the fact that the universe is expanding rapidly from the Big Bang – the hot, dense state in which it originated. According to the best

Inflation: proof for a multiverse?

1 Big Bang
In an infinitely dense moment some 13.8 billion years ago, our universe was born from a singularity.

2 Cosmic microwave background
After about 380,000 years, subatomic particles known as electrons cooled enough to combine with protons. The universe became transparent to light and the CMB began to shine.

3 Dark ages
Clouds of dark hydrogen gas cooled before joining together.

4 First stars
Gas clouds collapsed and fusion in stars began.

5 Galaxy formation
Gravity caused galaxies to form, merge and drift. Dark energy accelerated the expansion of the universe, but at a much slower rate.

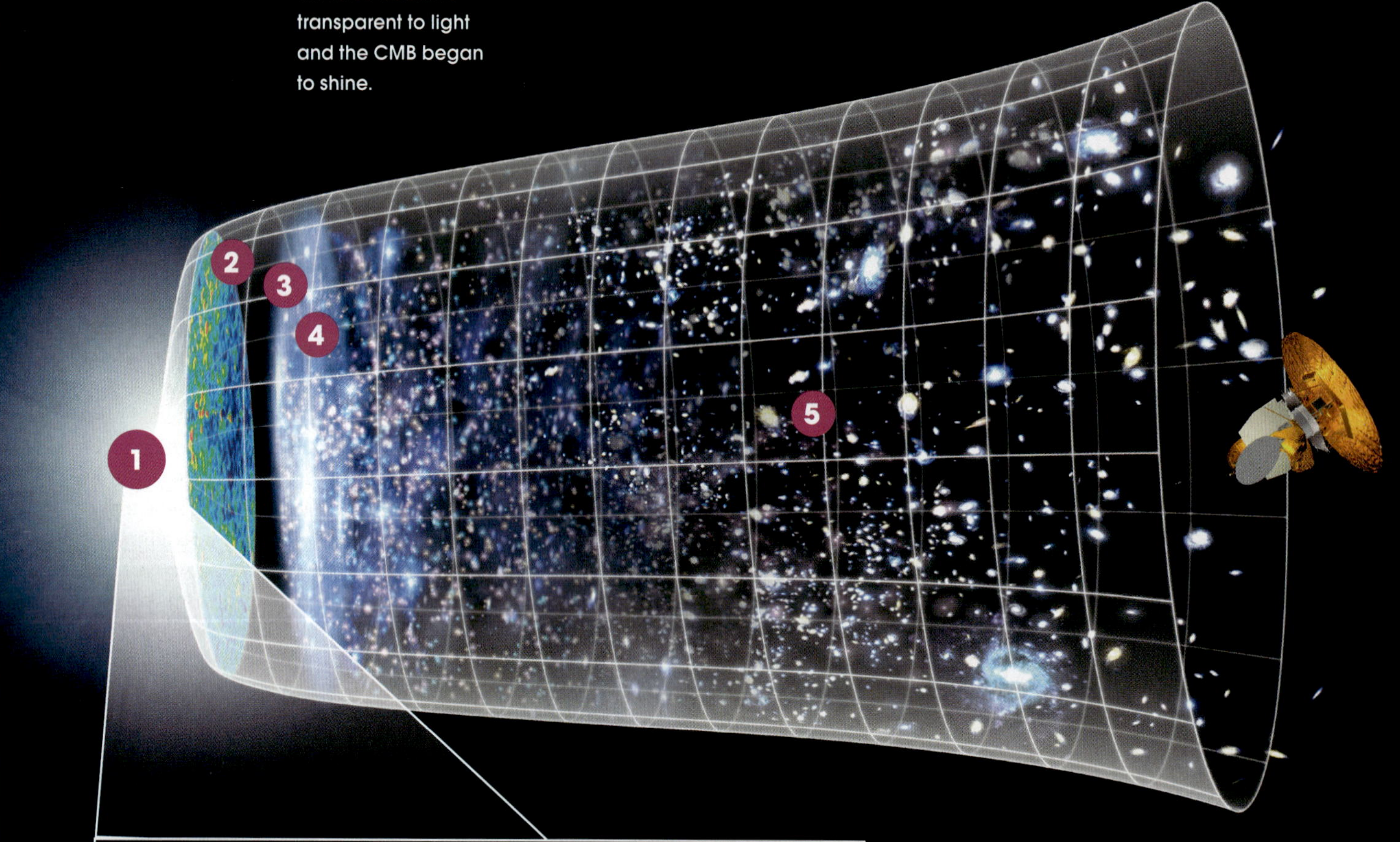

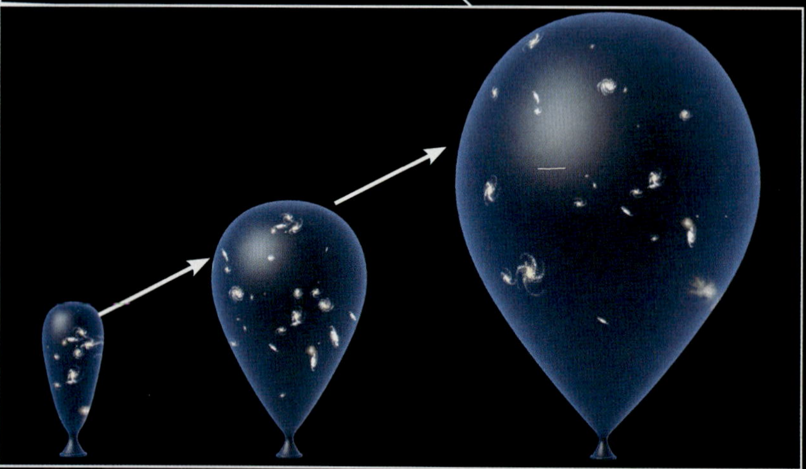

Rapid inflation

According to the theory of inflation, the universe expanded rapidly for a fraction of a second after the Big Bang. The easiest way to visualise the inflation of the cosmos is by blowing up a balloon and drawing galaxies on it – you will see the galaxies move away from each other as you blow air into the balloon.

current measurements, the Big Bang happened 13.8 billion years ago, and so we can only ever see objects whose light, travelling at 299,792 kilometres (186,282 miles) per second, has had time to reach us. Because the very early universe was so dense, it formed a brilliant opaque fireball that only became transparent after about 400,000 years. Light from that fireball – transformed into invisible microwave radiation in its long journey across expanding space – is the most distant thing we can directly observe. This cosmic microwave background (CMB) radiation has a key role to play in the search for evidence of a multiverse. Its true distance is estimated at 46.5 billion light years because, although the most distant light has been travelling for 13.8 billion years, the space it has been moving through has been expanding during that time.

"We don't know what's beyond the horizon," continues Kleban, "but what we can do is extrapolate from what we can see. On those large scales, the universe is pretty much homogeneous and isotropic, meaning it's pretty much the same in all directions, and as far as we can tell the same in every place. There's definitely a wider universe that's much bigger than what we can see, but it may not be very interesting. That's a basic assumption of modern cosmology that we call the cosmological principle."

There is growing evidence that we live in a multiverse, with multiple Big Bangs bringing other universes into existence

A professor of physics at New York University, Matthew Kleban believes that it's highly likely we live within a multiverse

Quite how far these distant reaches of space-time stretch is an intriguing question, and one that depends on the shape of space itself. Estimates range from about 250 times the size of the observable universe for a 'closed' and finite universe in which space curves inwards like a sphere, to infinite if space is flat or 'open', curving outwards like a saddle. However far space stretches, though, we'd expect parts of this wider universe to be essentially similar to our own. If the universe really is infinite, or close to it, we'd expect there to be parts of the universe, far away from us, that are exactly like our own observable universe.

How far away would these replica 'universes' – and the replicas of ourselves that would live in them – be? Our observable universe, with its radius of 46.5 billion light years, has enough room for 10118 particles. Try to imagine all the different ways these particles can be arranged – mathematics tells us that there are two to the 10118 different arrangements of all these particles and that we would have to cross ten to the 10118 metres – that's ten to the power of ten with 118 zeroes after it – before we encountered another duplicate universe with duplicate versions of us and our friends, living out parallel existences. That's a long way. In comparison, our observable universe is just 8.8 x 1026 metres across, but if the universe is infinitely big then there's enough room for the particular arrangement of particles making up our universe to be repeated again and again.

"There's definitely a wider universe that's much bigger than what we can see"
Matthew Kleban

What if there's another type of multiverse – one in which universes pop into existence like bubbles and have the potential to be radically different from our own? That's the intriguing possibility that fascinates Kleban and many of his colleagues, and at its heart lies a concept called eternal inflation. "In everyday life we're familiar with the phases of matter – if you think about a water molecule, for example, that can be liquid water, ice or steam. But in fundamental physics it's not just substances that have phases, but everything around us, space and time themselves. Theories like string

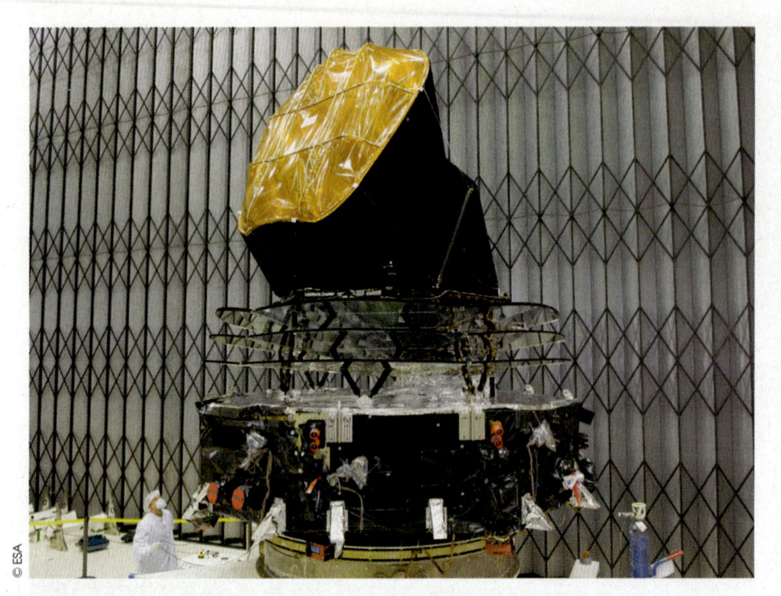

The Planck telescope being prepared for tests at the European Space Research and Technology Centre (ESTEC) in the Netherlands

"If the universe is infinite, we'd expect there to be parts of it that are exactly like our own observable universe"

theory predict a large number of phases, differing a lot more than ice differs from water. They would have different laws of physics. For instance, the fundamental particles of the cosmos, such as electrons and quarks, might not exist in some other phase. They might have some different form, or different properties like electrical charge and mass. These phases are kind of a generic feature of most modern cosmological theories."

Among the various properties that could change from one phase to another is the strength of the 'vacuum energy' that permeates empty space. In the past couple of decades, astronomers have discovered strong evidence that a small amount of this energy in our universe – better known as 'dark energy' – drives the expansion of the cosmos to accelerate when it should be slowing down, but in other phases it might be nonexistent, much stronger or even have a negative value. This, it turns out, is key to creating new universes in this kind of multiverse. If all these different phases can exist, there should be transitions between them, just as there are between ice, water and steam.

"You could have a universe that at some initial time has a single phase everywhere," Kleban explains, "but bubbles of different phases will inevitably appear more or less at

The different types of multiverse

From unseen regions of space-time to complex structures, there are four distinct levels

Level 1

Where an identical Earth exists

The simplest multiverse is one that most certainly exists – the Big Bang model of cosmic origins predicts that every point in the universe has a 'Hubble volume' around it, limited by the expansion of the universe and the distance light has been able to travel in the 13.8 billion years since the infant universe became transparent. In practice, this means our Hubble volume is a sphere 93 billion light years across, but there are many more Hubble volumes extending far beyond what we can see. If the universe has a closed geometry, the number is limited as space curves back around on itself, but if the universe is open, as seems most likely, there may be an infinite number of Hubble volumes, each containing a universe, meaning that somewhere out there, other planets virtually identical to Earth exist.

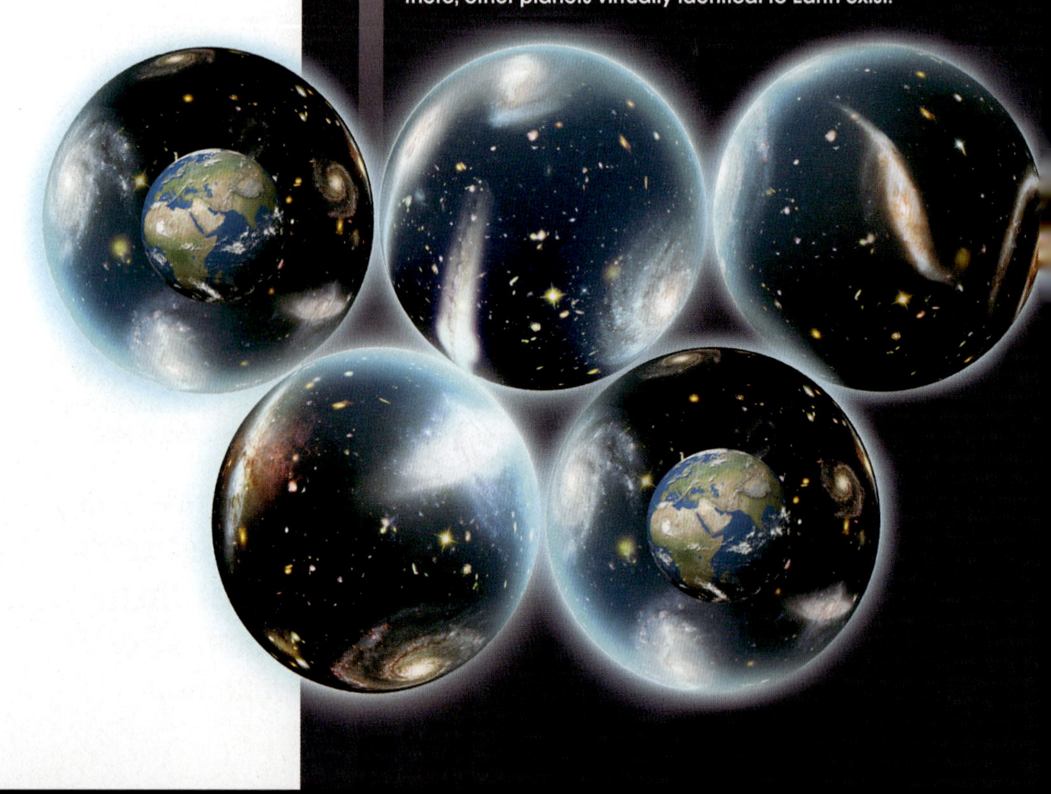

Level 2

The expanding universe we can't reach

String theory, a potential grand unified theory that aims to explain the fundamental laws of particle physics, suggests that space-time has at least ten dimensions, of which we experience just four – three dimensions of space, plus time – in our universe. The others are tightly curled around each other so we don't perceive them. But our arrangement of dimensions, or phase, is just one among many possible phases – given the right conditions, new ones can pop into existence within a pre-existing phase and then expand rapidly at the speed of light, meaning that they are completely unreachable. This theory of eternal inflation gives rise to a potentially infinite number of bubble universes with different dimensions and laws of physics.

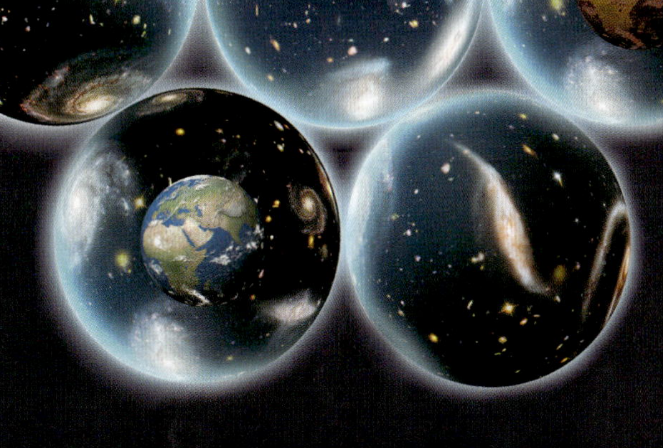

Level 3

Where your future self exists

According to the many-worlds interpretation of quantum mechanics, every decision point between outcomes – even on the tiniest microscopic scale – sees the universe branch into two mutually unobservable realities. This would involve the creation of multiple universes not in the space described by string theory, but in an infinite-dimensional geometric structure called a Hilbert space. Although you might imagine the many-worlds interpretation giving rise to a more varied multiverse than an extended or bubble model, the reality is that since all three are infinite, the same variety will play out. Some physicists have even argued that if quantum mechanics works in a certain way, then the many-worlds version of the multiverse could be formally equivalent to the more mundane 'single-geometry' versions.

Level 4

The strangest universe of all

As if the ideas of a multiverse as an extension to our own universe, a series of interconnected bubbles or a branching structure of infinite dimensions weren't strange enough, cosmologist Max Tegmark of the Massachusetts Institute of Technology argues that all of these 'lower level' multiverses are simply limited examples of an overarching mathematical multiverse called the 'ultimate ensemble'. This encompasses all possible multiverses that can be abstracted from the messy details of terminology into a purely mathematical description, and therefore includes all the lower level multiverses, plus any other types that still remain to be discovered. This, however, is merely part of Tegmark's semi-philosophical argument that the entire multiverse is a mathematical structure within which conscious entities perceive a physically 'real' world.

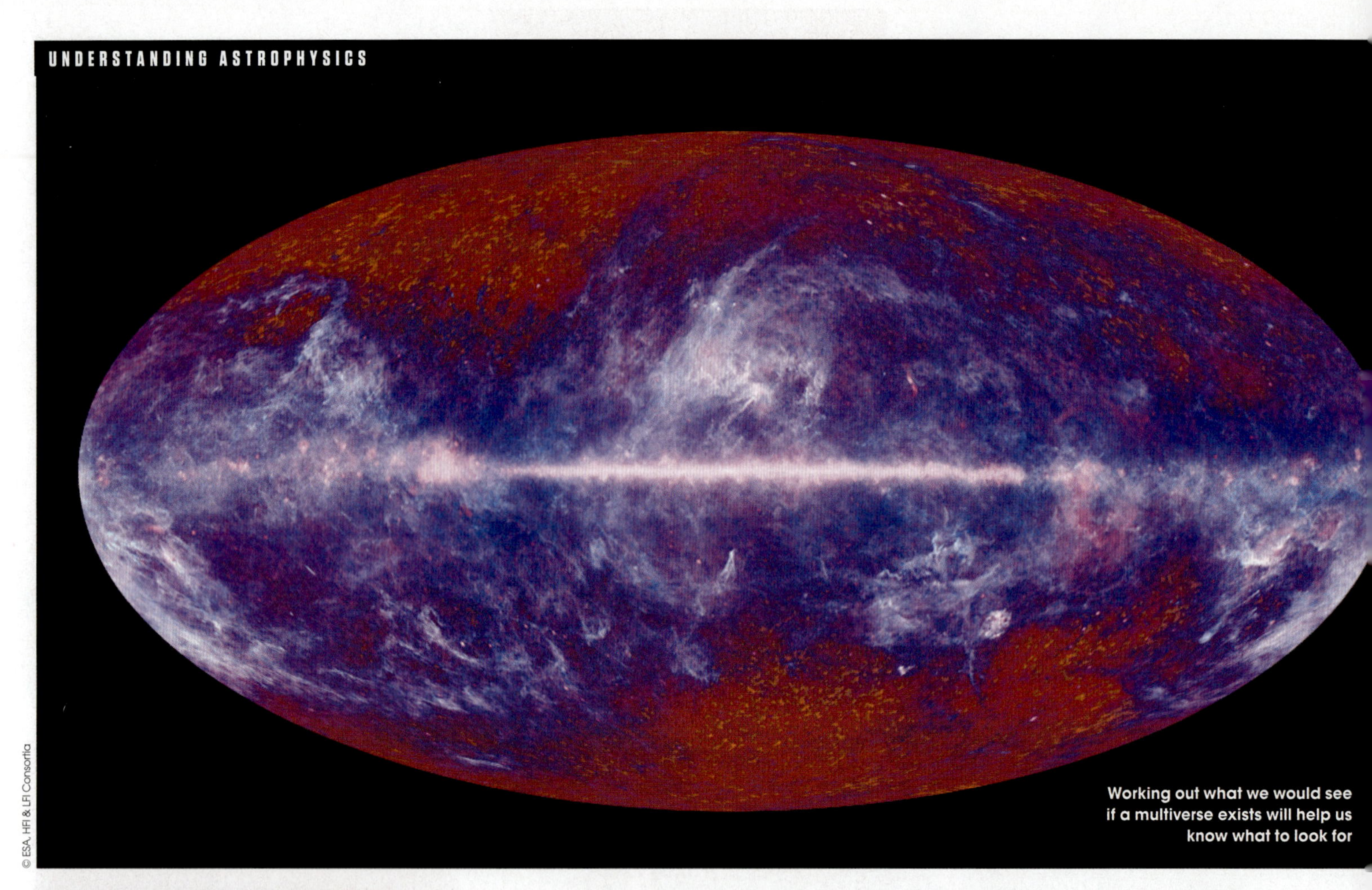

Working out what we would see if a multiverse exists will help us know what to look for

Data from the European Space Agency's defunct Planck satellite is being used in an attempt to prove multiverse theory

> **"There are some anomalies in the CMB that look a little bit like bubble collisions"**
> **Matthew Kleban**

random, like bubbles appearing in champagne. It's sort of a coin flip whether a given phase would have positive or negative vacuum energy, but at least some of them will be positive. If the vacuum energy is large, then the bubble expands exponentially, doubling in a fraction of a second, doubling again after that and so on. The volume will just explode in those regions. And if those phases are themselves unstable, then bubbles will appear inside them – that's what we call eternal inflation." It's a fairly mind-blowing concept, as Kleban admits: "If all this is correct, we may be inside one of these bubbles, and outside of it is something that's probably extremely exotic, most likely very rapidly inflating and has different laws of physics – perhaps even different numbers of dimensions. Once you go beyond the wall of our bubble, the multiverse isn't at all boring and isotropic after all."

Eternal inflation would explain one of the biggest mysteries about our cosmos. The properties of the universe seem suspiciously fine-tuned for life. If the gravitational constant were a little stronger, the charge of an electron a little smaller or the force that binds particles together weaker, then stars and planets would not be able to form and we would not be here. Everything is 'just right', like Goldilocks' porridge, and so far no one has been able to explain why. However, if there are an infinite number of bubble universes, all with slightly different properties, then there's bound to be a universe – our own – where the properties are just right, which would explain why we can exist.

One of the big questions is whether we could ever hope to find evidence that would prove the theory – or at worst prove that this cannot be the case. The idea that the multiverse theory cannot be proved or disproved has been a common criticism from sceptics, and it's an area where Kleban has been concentrating much of his work and research. "What's nice about the theory is there are observational consequences – if other bubbles form close enough to us, then they'll collide with our own bubble. Detecting the consequences with our current technology is a long shot, but it's not impossible, and you can actually work out what you would see, and therefore know what to look for," explains Kleban. "The theory makes testable predictions, and it also makes falsifiable predictions. It predicts that our bubble would need to have an open spatial geometry – if we measure the geometry of our universe and it turns out to be closed, then that would falsify the whole thing."

So what traces would a collision with another bubble make on our universe? As you might expect, a collision between two universes is a highly energetic event. "The walls of these bubbles are extremely rigid, and moving at speeds very close to the speed of light because there's a force that drives their expansion," enthuses Kleban. "The bubble naturally wants to expand and 'eat up' the vacuum energy around it. That gets converted into the kinetic energy of the wall, so these things accelerate, going faster and faster, until they collide. The result is a wave of energy injected into our own bubble, and this propagates across our universe in what we call a 'cosmic wake'. All sorts of things are affected, but the most important is the cosmic microwave background radiation. That's what we want to look at, because it's the oldest and most distant radiation, so it's had the most time to be affected by this sort of event."

According to most simulations, a collision between universes would show up as a ring of slightly higher temperature amid the otherwise random variation of the CMB, and also as a polarisation pattern within the radiation – instead of microwaves vibrating in random planes, their oscillations would be aligned in specific directions. "So far we haven't found any strong evidence for such collisions, but there are some anomalies in the CMB that look a little bit like bubble collisions. I don't take them very seriously, but imagine there really was something there on the verge of detectability. It would produce an anomaly of marginal significance. We're waiting on a major new piece of data in the form of an all-sky polarisation map from Planck. That's a partially independent piece of data from the temperature maps, and certain types of bubble collisions would have a very distinct signature in the polarisation."

If our universe does turn out to be just one among infinitely many in a multiverse, the implications for cosmology would be huge. No longer would we imagine space and time as being created in a single event 13.8 billion years ago – that would simply mark the point when our particular phase popped into existence and began to expand. But as Kleban is keen to point out, the study of our potential multiverse is still very much in its infancy. "We focus on certain types of collisions because it seems like those would be the ones we'd have the best chance of observing, but there could be something that we have missed. And the other possibility is an entirely different type of observation that we can't yet do, or haven't thought of. The important thing is that it is possible to detect the multiverse, and once something's possible, we may discover a clever way to do it. We're really still at the beginning."

Giles Sparrow
Space science writer

The author of over 20 books on popular science, Giles holds a degree in astronomy and is an editor specialising in science and technology.

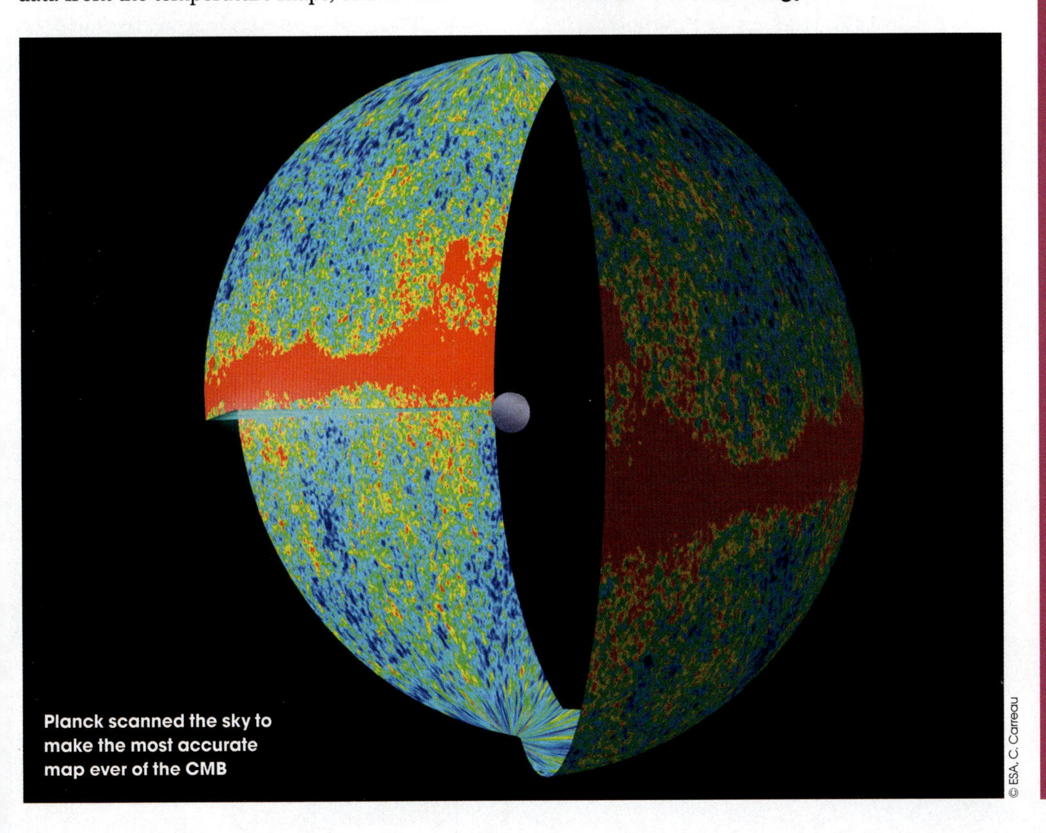

Planck scanned the sky to make the most accurate map ever of the CMB

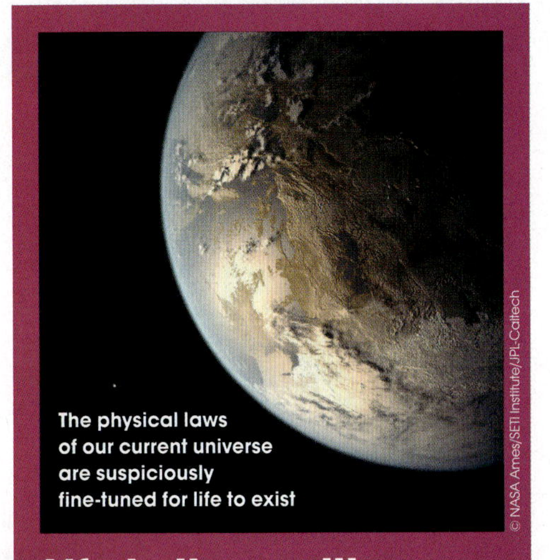

The physical laws of our current universe are suspiciously fine-tuned for life to exist

Life in the multiverse

Confirmation of a multiverse could raise some intriguing questions about the origins of life and our own place in the cosmos. According to our current understanding, the physical laws of our universe are suspiciously fine-tuned for life to arise – a tiny difference in any one of several physical constants might make liquid water much rarer, render complex organic chemistry unworkable or leave matter itself unable to hold together. The usual scientific solution to this problem, called the 'weak anthropic principle', simply states that if the laws of the universe didn't have the particular behaviours we observe, we wouldn't be around to see them, so we shouldn't be so surprised. 'Strong anthropic principles' take things a step further with the assumption that, for some reason, the universe has to give rise to something like its present set of life-friendly parameters, perhaps even because the existence of conscious observers is a requirement for the universe to exist.

If this universe is just one within an infinite multiverse, this could significantly change the terms of the debate. The odds of a universe arising with our own parameters would rise from highly improbable to a locked-in certainty, but at the same time so would the odds of any other combination of parameters. So could life exist in these other universes? According to some researchers, it may be far more robust than previously realised. Using computer simulations to study the evolution of universes with various fundamental constants, they have found that stable forms of matter and carbon chemistry can arise in a surprising variety of situations.

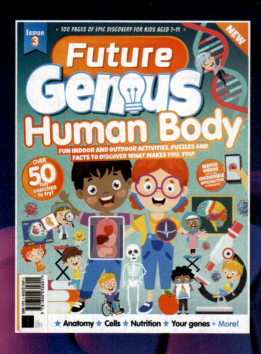

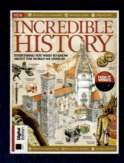

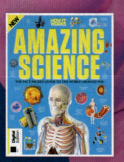

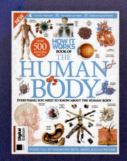